THE

SCIENCE

OF

SUPERWOMEN

THE
SCIENCE
OF
SUPERWOMEN

AN EVOLUTION FROM *WONDER WOMAN* TO *WANDAVISION*

MARK BRAKE
AUTHOR OF *THE SCIENCE OF SUPERHEROES*

Skyhorse Publishing

Skyhorse Publishing books may be purchased in bulk at special discounts for sales promotion, corporate gifts, fund-raising, or educational purposes. Special editions can also be created to specifications. For details, contact the Special Sales Department, Skyhorse Publishing, 307 West 36th Street, 11th Floor, New York, NY 10018 or info@skyhorsepublishing.com.

Skyhorse® and Skyhorse Publishing® are registered trademarks of Skyhorse Publishing, Inc.®, a Delaware corporation.

Visit our website at www.skyhorsepublishing.com.

10 9 8 7 6 5 4 3 2 1

Library of Congress Cataloging-in-Publication Data is available on file.

Cover design by David Ter-Avanesyan
Cover photo by Shutterstock.com

Print ISBN: 978-1-5107-7631-9
Ebook ISBN: 978-1-5107-7632-6

Printed in the United States of America

This book is dedicated to Amelia, Bryher and Bryony, Ellie and Emily, Frances and Helen, Margaret and Megan, and last but never least, Rosi; warriors and superwomen, all

TABLE OF CONTENTS

INTRODUCTION

The superhero genre is a billion-dollar business. By the middle of the second decade of the twenty-first century, there were more than two dozen superhero television programs being broadcast or in planning, and more than fifty movies showcasing superhero characters. This media operation is across all platforms: the original comics, movies, animated and live-action TV programs, and fan conventions—all with linked and licensed merchandise. Transmedia gone global.

According to Statista, the German online global data and business intelligence platform, at the time of writing, the total global box office revenue for Marvel Cinematic Universe (MCU) movies is $29.55 billion, the total global box office revenue for DC Extended Universe movies is $6.59 billion, and the annual box office revenue of superhero movies in the US and Canada from 1978 to 2021 is $1.44 billion. Yet, of these media commodities, only around 7 percent are solo vehicles for superwomen. Of the remainder, around half have white male leads, and the rest have near total male (and white) ensemble casts. It's a similar story in mainstream superhero comics, where the percentage starring female superhero characters is at an all-time high of only 12 percent.

The multimedia superhero genre in film, TV, and comic books stands as a very prominent cultural space that underrepresents women in places of power. This is true both in terms of fictional characters *and* as the real-life creators of those characters. So, the question of the very nature of the characters and stories of superwomen inspires huge passions in which fans feel emotionally invested.

This book explores the corporate productions, creator representations, and audience receptions of female superheroes in mainstream superhero comics, television shows, and films. It examines how and why they're created and sold the way they are, how and why they sound and look the way they do, and how and why different audiences make meaning from them.

As the highest-grossing film franchise of all time, MCU movies have grossed more than twice the takings of *Star Wars*, with the average MCU movie grossing a billion dollars apiece. We shall see in this book that superhero fiction has been with us for almost a century; high-octane tales crammed with concepts and contrasting themes from superpowers and the post-human to masked vigilantes and immortality. In that time, superwomen have evolved from comic book caricatures (created by men, for men) to better representations of female power.

This book looks at that evolution, from its hypersexualized origins to today's attempts at a more nuanced diversity. We explore the relationship between superhero film and fiction and the underlying current of our ever-evolving planet. We begin in the nineteenth century with the rather quirky Chapter 1: How Did Superwomen Lead to the First Sci-Fi Franchise? (page 1) telling the tale of subterranean superwomen and their links to a hot beverage. In Chapter 2: What Ideas of Hell Are in *Buffy the Vampire Slayer*? (page 11), we explore the story of one of television's greatest ever superwomen and her relationship with hell and damnation. Chapter 3: Human and Monster: What Super-Females Are in *Alien*? (page 27) considers the female hero of a sci-fi/horror classic and her monstrous female nemesis of a xenomorph. Chapter 4:

How Has Wonder Woman Changed with the Times? (page 43) grapples with arguably the greatest ever comic superwoman, as we look at the social, psychological, and political science of Diana Prince. In Chapter 5, Why Is Shuri Such a Big Deal? (page 79), we look at the various social, cultural, and educational factors in the public's perceptions of cinema scientists. In Chapter 6: How Did Super-Barbie Go from Plastic to Profound? (page 107), we peel back the feminist layers of Greta Gerwig's fabulous film, and in Chapter 7: Is Storm Wise to Tinker with the Weather? (page 119), we mull over the consequences of messing with chaos theory. Chapter 8: What of Witches, Wednesday Addams, and WandaVision? (page 127) brings together a brief history of witches alongside two very contemporary screen "witches" in Wednesday Addams and the Scarlet Witch, and finally in Chapter 9: Are Female Superheroes Less Profitable? (page 149) we lift the lid on superheroes and corporate power.

CHAPTER 1

HOW DID SUPERWOMEN LEAD TO THE FIRST SCI-FI FRANCHISE?

(. . . in which we tell the tale of subterranean superwomen and a hot beverage)

"7 April 1852/Went to the zoo/I said to him—Something about that chimpanzee over there reminds me of you."
—Carol Ann Duffy, Mrs. Darwin in *The World's Wife* (1999)

MRS. DARWIN

Carol Ann Duffy is a Scottish poet and playwright. For a decade, between 2009 and 2019, she was poet laureate of the United Kingdom. She was the first female poet, the first Scottish-born poet, and the first openly gay poet to hold the poet laureate position. In 1999, Duffy published *The World's Wife*, a collection of poems written from the perspectives of the wives of various famous and fictional men.

One of the poems, "Mrs. Darwin," starts on a specific day in 1852, around seven years before Darwin's *On the Origin of Species* was published. The speaker, Charles Darwin's wife, pithily recalls a trip to the zoo with her husband. In particular, her pointing out a nearby chimp and suggesting the creature bore an uncanny resemblance to her husband. Duffy's amusing implication is that Darwin's historic theory, that human beings and apes have a common ancestor, was in fact sparked by *Mrs.* Darwin. Given that a woman was really behind this world-changing idea, Duffy's poem makes the reader wonder how many more women's contributions to history have been overlooked by society. Last, the poem subtly subverts Darwin's stature, a poster boy for the science establishment, reminding readers that he's as apelike (and human) as everybody else.

FOSSIL HUNTERS

The idea of evolution had been in the air for some time. British paleontologist, Mary Anning, was born a decade before Darwin. She dedicated her life to fossil collecting from the age of twelve, and once helped her father dig up an ichthyosaur skull. Her family lived nearby the cliffs along the English Channel at Lyme Regis in the county of Dorset in Southwest England. Mary's discoveries in those Jurassic marine fossil beds became known around the world for helping paint a picture of life in the past and lending important evidence to the idea of change in species over time. (Mary's life and legacy were made into a movie, *Ammonite*, that premiered at the Toronto International Film Festival on September 11, 2020, in which actress Kate Winslet portrayed Mary.)

Even though Mary knew more about fossils and geology than many of the wealthy male collectors to whom she sold, it was always the gentlemen geologists who published the scientific descriptions of her specimens, often neglecting to mention Mary's name. Naturally, Mary became resentful of this. Mary was cursed by her social class, as well as her sex. The slights to Mary were part of a trend of disregarding the contributions of working-class people in nineteenth-century scientific literature. Many fossils would be found by quarrymen, construction workers, or road workers who would sell them on to a wealthy collector, and the latter was credited if the fossil was of scientific import.

INCENDIARY EVOLUTIONISTS

All this fossil evidence had to lead somewhere, as the idea of evolution had long been in the air. Yet you'd not know it if you witnessed "Darwin year," or the bicentennial of Darwin's birth, in full flight. It was 2009. All across the globe, but above all in Britain, programs of celebration were organized to commemorate Charles Darwin's life, times, scientific ideas, and their impact. Many organizations and agencies, large and small, were involved. They ranged from arts councils and the BBC, through Westminster and the Woodland Trust, to various universities and zoological societies up and down the land. They planned and executed an incredible array of events to honor the life and legacy of a *single* biologist: conferences and exhibitions, graphic novels and dramatic performances, and documentaries on radio and television.

Darwin was big business. And, in 2009, the doctrine of Darwinism was on display. The Natural History Museum in London held a Darwin Big Idea Exhibition. Billed as the biggest ever exhibition about Charles Darwin, it celebrated "the impact of the revolutionary theory that changed our understanding of the world." Regrettably, the exhibition's campaign poster appealed to the faithful with the absurd words "If you had an idea that was going to outrage society, would you keep it to yourself?"

Now we can more clearly see the point Carol Ann Duffy was making in "Mrs. Darwin." The theory of evolution was not the work of a single male "genius," keeping secret an idea that would "outrage society." A stream of very able thinkers, running from Empedocles, Epicurus, and Lucretius to Leonardo da Vinci, had tended toward a secular speculation at the rich variety of life on Earth. So too did luminaries such as Francis Bacon, René Descartes, and Gottfried Wilhelm Leibniz (and possibly their wives!), all of whom favored natural over supernatural causes for species change. And the French naturalist Étienne Geoffroy Saint-Hilaire had made significant progress in the late eighteenth century, suggesting that some species adapt and survive when environments change.

ZOOMANIA

"Mrs. Darwin" aside, Charles Darwin was blessed with a familial vision that proved hard to live down. His grandfather, Erasmus Darwin, was an ingenious mechanic, inventing a speaking machine, a mechanical ferry, and a rocket motor. He was also a provocative evolutionist. As a boy, Charles had poured over Erasmus's mighty work on evolution, *Zoomania*, published in two volumes in 1794 and 1796. It was replete with hearty exclamations that life had evolved from a single ancestor.

Erasmus Darwin was a celebrated communicator of science. Romantic poet Samuel Taylor Coleridge declared him "the first literary character of Europe, and the most original-minded man." One of Erasmus's poems on evolution enjoys a science-fictional vision. It foresees, with unerring accuracy, a future of colossal skyscraper cities, overpopulation, convoys of nuclear submarines, and the advent of the car. Young Charles didn't inherit his grandfather's boldness of spirit. Science historian A. N. Whitehead declared that "Darwin is truly great, but he is the dullest great man I can think of." Maybe Carol Ann Duffy was right about the zoo! Constrained by tradition and church, Darwin was reluctant to publish about evolution for many years. The blow to religion and the social order struck by this theory would inspire "atheistic agitators and social revolutionaries." Evolution should have been countenanced long before. Opposition from landed and clerical interests, however, feared its deadly threat to the divine ordering of the world.

Evolutionary theory struck at the heart of what it was to be human. Isaac Newton's system of the world had essentially reestablished the integrity of divine design, which had been shattered by earlier discoveries in astronomy. But the divine picture of creation had stayed more or less untouched. Humans were still made in the "image of God." After evolutionary theory, the book of Genesis lay in shreds as a literal history. Darwin's *Origin of Species*, published in 1859, was appropriated by the radical, anticlerical wing in politics, and molded to its agenda of laissez-faire capitalism. Evolutionary theory provided an alibi for brutal exploitation by the "fittest," the subjugation of lesser by higher peoples. Association with nature "red in tooth and claw," as Tennyson put it, could justify even war itself. And the notion of the "chosen ones," that one-time apology for the supremacy of classes or races, had withered. It was replaced by a "Darwinian" validation of a brave new world of reason, industry, and empire.

The idea of evolution also injected the lifeblood of history into science. "He who . . . does not admit how vast have been the past periods of time may at once close this volume," Darwin wrote in the *Origin*. Evolutionary theory *could* have been used to unite the human and nonhuman spheres, but the social evolution of humanity was eclipsed by scientism. A limited and penetrative science focus resulted in the perverse justification of race theories and imperialism. And utopian tales of the future that explored the connection between nature and society came to dominate.

A MILLION, MILLION SUNS

In this social atmosphere, the previous idyllic vision of a static world became passé. In its place was mutability. The utopian tales stressed the ebb and flow of evolution as a reaction to the unsettling changes in the fabric of Victorian society. In short, after Darwin, the new paradigm was the process of becoming; the question as to what would become of humankind. In the words of Tennyson, "Earth's pale history runs, What is it all but a trouble of ants in the gleam of a million, million suns?"

During the development of the nineteenth century, early progress gave way to pessimism. The economic slump of the Great Depression in Britain between 1873 and 1896 marked the end of unquestioned expansion. In 1870 the country had been liberal, but by the middle of the 1890s British politics was sharply polarized and most of its capitalists had seceded to the conservatives. The first cloth-capped proletarian socialist sat in Parliament and new utopian fiction arose as the social order began to corrode, with many a narrative offering a parody on the disorder of the contemporary culture.

Science was not wholly responsible for the Great Depression and its ills. It was correctly felt, however, that science had transformed industry. This revolution cultivated the urban, class-conscious culture of industrial Britain. In addition, the radical displacement of Christian faith brought about by Darwin had ironically led to an alienation from nature. As the problems of the age grew more complex and challenging, historians of the future delivered bleaker forecasts.

THE POWER OF THE COMING RACE

This scientific and political context inspired one of the first evolutionary fables about superwomen, Edward Bulwer-Lytton's 1871 novel, *The Coming Race*. Edward George Earle Bulwer-Lytton, 1st Baron Lytton, politician and novelist, was friend to creator of the modern British conservative party, Benjamin Disraeli.

Lytton wasn't a happy man. He wasn't happy about the new ideas of evolution. And he was even unhappier about the ideals of John Stuart Mill in his book, *The Emancipation of Women*, so it was Lytton's intention to satirize both Darwinian biology and female emancipation. *The Coming Race* includes a pseudoscientific account of an evolved line of humans who believe they are descended not from apes but from frogs. Lytton's satire also features a somewhat inelegant gender reversal. The women are fitter, beefier, more assertive, and hairier than the men.

His fascinating, if bizarrely paranoid, tale is set in a subterranean world of well-lit caverns. It begins as the narrator, an American mining engineer, falls into an underground hollow. There he discovers a mysterious human-like race, the Vril-ya. These humanoids derive immense power from *vril*, an electromagnetic animating force which fuels air boats, mechanical wings, formidable weapons, and automata:

> In all service, whether in or out of doors, they make great use of automaton figures, which are so ingenious, and so pliant to the operations of *vril*, that they actually seem gifted with reason. It was scarcely possible to distinguish the figures I beheld, apparently guiding or superintending the rapid movements of vast engines, from human forms endowed with thought.

The scientific dream of an automated society had been realized by this race of subterraneans. The unEarthing of *vril*, the "all permeating fluid," borne by strident emancipated females, had enabled the race to master nature. Gender equality had been more than achieved. War had been eliminated through mutually assured destruction. In Lytton's words, it was a utopia that made real "the dreams of our most sanguine philanthropists."

But Lytton rejects this utopia, this "angelical" social order. The sociable community of the Vril-ya has eliminated competition, but is, according to Lytton, barren of those "individual examples of human greatness, which adorn the annals of the upper world." He means *male* greatness, of course. Not only that, but Lytton thinks suffering is a good idea, as long as it's not actually Lytton doing the suffering, naturally. According to him, conflict and competition, misery and madness, all are innately "human." Sounding a note struck more clearly in Huxley's *Brave New World*, Lytton brands female emancipation and "calm and innocent felicity" as vain dreams. Utopia, and the displacement of human industry to *vril*, would lead only to enervation and ennui.

"OUR INEVITABLE DESTROYERS"

The Coming Race is a key marker of the Victorian obsession with evolving society. While a new kind of life is secured through the application of science, the novel suffers from machine determinism. The harnessing of technology made the Vril-ya formidable, but the new technology has no social agency (even though it is described as having inevitable social consequences). Far-flung subterraneans who do not have *vril* are uncivilized (or, we might say, "the great unwashed," since Lytton is alleged to have coined the phrase). Indeed, the possession of *vril* energy *is* the civilization, and the refinement of society is based on technology alone.

The book strikes a fearful note at the prospect of the rise of superwomen. As suggested by the ominous title, once the more advanced Vril-ya surface from their caverns, they will take the place of mere men: "the more deeply I pray that ages may yet elapse before there emerge into sunlight our inevitable destroyers."

Meanwhile, back in the real world, white women over the age of twenty-one were allowed to vote in the western territories of Wyoming from 1869 and in Utah from 1870. And New Zealand became the first self-governing country to grant all women the right to vote in 1893, when women over the age of twenty-one were permitted to vote in all parliamentary elections. The coming race was real.

A MOST UNEXPECTED FRANCHISE

Finally, that superwomen franchise. A modern media franchise is a collection of related and derivative products inspired by an original creative work of fiction or film. Media franchises tend to cross over from their original form into other markets. Literary franchises are often transported to film, such as Miss Marple, Nancy Drew, other popular detectives, and, of course, popular comic book superheroes: Wednesday Addams Funko POP! figures, Wonder Woman cosplay costumes, Buffy the Vampire Slayer felt figures, that kind of thing.

Edward Bulwer-Lytton's *The Coming Race* can lay claim to being the first ever franchise product of the sci-fi genre. Indeed, when H. G. Wells's *The Time Machine* was published in 1895, *The Guardian* said in its review:

> The influence of the author of *The Coming Race* is still powerful, and no year passes without the appearance of stories which describe the manners and customs of peoples in imaginary worlds, sometimes in the stars above, sometimes in the heart of unknown continents in Australia or at the Pole, and sometimes below the waters under the Earth. The latest effort in this class of fiction is *The Time Machine*, by H. G. Wells.

The Coming Race was also the inspiration for the fortune made from Bovril, a thick and salty meat extract paste developed in the 1870s by Scottish entrepreneur John Lawson Johnston. For skeptical readers in

North America, Bovril can be made into a drink, sometimes known in the UK as a "beef tea," by diluting with hot water. The first part of the name, *bovis*, comes from the Latin meaning for ox, and the second part, *vril*, comes from Lytton's novel. This drink is clearly meant to derive the great strength of an ox, in the same way that superwomen derived their powers from the electromagnetic substance of *vril*.

An early prototype of Bovril was provided by Johnston's company to Napoleon III, who had requested one million cans of beef to feed his troops during the Franco-Prussian War. By 1888 in the UK, over three thousand public houses, grocers, and dispensing chemists were selling Bovril, and by the early twentieth century Bovril was being promoted as a superfood that could protect one from influenza. And Bovril holds the curious distinction of having been advertised with a Pope. An advertising campaign in Britain in 1900 showed a throned Pope Leo XIII holding a mug of Bovril. The advertising slogan read "The Two Infallible Powers—The Pope & Bovril."

There we have it. The most unlikely story of science fiction's first franchise. It began with the pen of a paranoid Tory. And it was ultimately appropriated by the patriarchy to such an extent that they appointed an "infallible" Pope as "guarantor." And all based on the terrifying prospect of a future society of superwomen.

CHAPTER 2

WHAT IDEAS OF HELL ARE IN BUFFY THE VAMPIRE SLAYER?

(. . . the story of one of television's greatest ever superwomen and her relationship with hell)

"The first thing I ever thought of when I thought of *Buffy: The Movie* was the little blonde girl who goes into a dark alley and gets killed in every horror movie. The idea of Buffy was to subvert that idea, that image, and create someone who was a hero where she had always been a victim. That element of surprise, of genre-busting, is very much at the heart of both the movie and the series."

—Joss Whedon, *Welcome to the Hellmouth*, DVD commentary (2006)

BUFFY THE VAMPIRE SLAYER

Before writing and directing Marvel's *The Avengers* (2012) and its sequel, *Avengers: Age of Ultron* (2015), and cowriting DC's *Justice*

League (2017), Joss Whedon's primary superhero character was Buffy the Vampire Slayer. In his first created television series, *Buffy the Vampire Slayer* tells the tale of Buffy Summers, the latest in a line of young women called to battle against the forces of darkness, including demons, vampires (naturally), and a bestiary of other monsters. Whedon's creation was a reaction to the tired Hollywood formula of the victimized blonde girls who, Red Riding Hood–like, wander off their chosen path and inevitably get murdered, in *every* movie. Instead, Whedon wanted to subvert the genre and create a female superhero. As Whedon has commented, this superwoman idea was present from "the very first mission statement of the show, which was the joy of female power: having it, using it, sharing it."

Buffy the Vampire Slayer soon became critically and popularly acclaimed, drawing in average audiences of between four and six million viewers on original airings, even though the series was broadcast on the relatively new and smaller WB Television Network. The resounding hit of *Buffy* meant a myriad of spin-offs, including books, comics, and video games. *Buffy* also drew in a lot of appreciation in fan film form, as well as in parody and academia. As a result, *Buffy* is thought to be a critical actor in the new wave of television series during the late 1990s and early 2000s which featured other notable strong female characters, including *Xena: Warrior Princess*, *La Femme Nikita*, and *Dark Angel*.

WELCOME TO THE BUFFYVERSE

Hardly surprising, then, that the Buffy franchise has been labeled the Buffyverse or Slayerverse. The Buffyverse refers to the fictional Universe centered around *Buffy the Vampire Slayer*, a place where supernatural phenomena exist, and supernatural evil can be countered by good actors fit for the fight against such forces. The existence of supernatural phenomena in the Buffyverse is what distinguishes Buffy's realm from the real world, though only a small number of humans are aware of this distinction. Moreover, many traits of the Buffyverse are characterized as good or evil and storylines reflect this, though certain narratives follow more paradoxically gray paths.

Whedon's subversion of the genre starts from the get-go. Episode one. It's the dead of night. The camera pans across the front entrance of Sunnydale High School to a backing track which sounds like a musical buzzsaw; there may be trouble ahead. We move inside and through the deserted corridors of the school until we stop at the darkened science lab, whose arcane apparatus looks more nineteenth than twentieth century with a brass microscope, a skull in a bell jar, and dark specimens under glass. The sudden sound of a shattering window unsettles us further. A young man and his pretty blonde date break into the empty building for some extracurricular activities. The blonde seems brittle and timorous, starting at the slightest sound, seemingly uneasy that devilry is lurking in the shadowy lab. The lad has all the swagger and charm of the arrogant young male in the movies, sneering at the girl's fears, and insisting the couple are quite alone. But hold on. What's this? The blonde's face instantly morphs into the form of a fanged and yellow-eyed demon, as she sinks her razored canines into the neck of her doomed date.

Welcome to the Buffyverse, where all is not what it seems. A weird and wonderful world, a realm where witches and vampires, hellmouths and magic, and demons and devils all exist, and the chance of the end of the world as we know it is a daily possibility. Mashing up sci-fi, horror, and high school melodrama, the main postulate of the Buffyverse is prosaic: "In every generation, there is a chosen one. She alone will stand against the vampires, the demons, and the forces of darkness. She is the slayer." Enter a "blonde bombshell" with a difference. The less-than-sweet sixteen-year-old Buffy Summers is expelled from her old high school in Los Angeles for torching the gym (it was jam-packed with bloody vampires) and shipped to the fictional town of Sunnydale, California, which just happens to sit atop a hellmouth, a portal which often yawns open between the picture-perfect world of Sunnydale and the dimension of hell with its legion of demons. Buffy's mission, should she choose to accept it, and which she does for the next seven years, is to keep the legion of demons at bay, as the hellmouth is a gravitational black hole of badness, drawing danger to Sunnydale and threatening an eruption of hell on planet Earth.

THE REALM OF THE BUFFYVERSE

Now, what, you might well ask, does this surreal and mystical world have to do with science? The answer lies in the understanding of both realms, the fictional Buffyverse and our own Universe, an understanding that revolves around the limits and boundaries of their laws. The magic of the Buffyverse seems wholly fictional at first; demons aren't real, after all. And yet, when you take a closer look, hidden in the logic of the Buffyverse is a similar science to our own. Each realm can be seen as a complex mechanism which runs according to a set of underlying fundamentals. And, when they are grappling with the supernatural, the writers on *Buffy the Vampire Slayer* must think through the ways in which the magic may, or may not, supersede our normal laws of nature.

A comparison with the wizarding world of *Harry Potter* may help here. The Potterverse is another magical Universe that must place limits on its fantastic feats to make the fiction more believable. J. K. Rowling is on record as suggesting that, though not explicitly stated in her books, wizards could not simply conjure money out of thin air. An economic system based on such a possibility would be grimly flawed and highly inflationary. Perhaps that's also why a limit was placed on Harry's use of the sorcerer's stone for alchemy. The stone's abilities were described as extremely rare and possessed by an owner who did not exploit its powers. And Dumbledore also says there is no spell that can bring people back from the dead. Sure, they can be reanimated into compliant beings on a living wizard's command, but they would be little more than soulless zombies with no will of their own.

THE CHEMISTRY OF THE BUFFYVERSE

As with the Potterverse, so it is with the Buffyverse. There are logical limits to the way a writer can create a magical Universe, just as there are limits to science. Witness to this fact is that magic in the Buffyverse is underpinned by a science of sorts. Consider the chemistry of the Buffyverse. As in the Potterverse, the synthesis of brews and potions is often carried out through concoctions derived from the ingredients found in the science lab.

In one tale, Buffy herself behaves like an analytical chemist. Her mother, Joyce, becomes ill, and Buffy is suspicious about the mysterious origin of the illness. She suspects foul play in the form of a spell. But the trace signature of a spell is usually invisible to mere humans, so Buffy concocts a counterspell of her own which enables her to see the "invisible" trace, in a similar way to that of chemists who use different wavelengths of light to detect the telltale presence of distinct "signatures" of chemical elements.

Buffy reads traces of spells in a similar way to cosmologists reading starlight. For example, both chemists and cosmologists know that the light leaving a star reveals much more about science than its mere place in the sky. The atoms of things that glow lead energetic lives. Their electrons are forever absorbing and emitting light. And if their surroundings are energetic enough, collisions between atoms let loose some or all of their electrons, enabling them to scatter light, so atoms leave their fingerprint signatures on the light being studied, which uniquely implicate the chemical elements that are responsible and present in the star. These signatures, in the form of black "absorption" lines, are what Richard Dawkins refers to as the "barcode of the stars." They provide evidence for some remarkable conclusions, including the cosmic abundance of the chemical elements, and the red shift of galaxies receding from the alleged Big Bang at the beginning of the Universe.

THE PHYSICS OF THE BUFFYVERSE

Then there's the physics of the Buffyverse. The writers of *Buffy* often draw on ideas from relativity and quantum mechanics to evolve innovative storylines. For instance, a schoolgirl becomes invisible after months of nobody noticing her, an interesting twist on the quantum notion that observation determines the outcome of a sub-atomic experiment. *Buffy the Vampire Slayer* also showcases plots that revolve around temporal tricks and time loops, teleporting demons and multidimensional portals. As Kip Thorne said in *The Science of Interstellar*, "speculations (often wild) about ill-understood physical laws and the Universe will spring from real science, from ideas that at least some 'respectable' scientists regard as possible."

Thus, the Buffyverse includes concepts similar to the hypothetical wormholes proposed by real-world physicists like Kip Thorne himself. And the exchange of ideas is a two-way process. In December of 2005, planetary scientists discovered that a small object in the Kuiper belt, a ring of icy bodies near Neptune, had a peculiarly tilted orbit. They named the object "Buffy" because, as with a lot of physics in the Buffyverse, the small object's orbit was not describable using standard scientific theories of how the Solar System formed.

THE BIOLOGY OF THE BUFFYVERSE

Given that *Buffy the Vampire Slayer* is a series brimming with monsters, it comes as little surprise that many of the monster ideas in *Buffy* are borrowed from real-world biology. We have demons that deliver death on their prey by jabbing them with poisonous toxins to paralyze them before they're wolfed down. Then there's the question of the biology of vampirism, which could be seen as a kind of contagion, a blood-borne disease caused by pathogenic microorganisms, like Hepatitis B (HBV), Hepatitis C (HCV), or Human Immunodeficiency Virus (HIV).

When primeval demons reemerge from "forever" slumber, their impact is akin to biological warfare. The show's plot often involves a demon infecting a host like some type of turbo-parasite, but one which kills the host so that the remaining shell may be used as an embryonic weapon of mass destruction, so it's no good trying to destroy the demon. That would simply result in spreading the warfare further, with thousands dying where once only one would have perished.

Indeed, in a 2001 paper, "Biological Warfare and the 'Buffy Paradigm,'" Anthony H. Cordesman of the Center for Strategic and International Studies in Washington, DC, argues that "the US must plan its Homeland defense policies and programs for a future in which there is no way to predict the weapon that will be used or the method chosen to deliver a weapon which can range from a small suicide attack by an American citizen to the covert delivery of a nuclear weapon by a foreign state." Cordesman's suggestion is to "think about biological warfare in terms of a TV show called *Buffy the Vampire Slayer* . . . think about the world of biological weapons in terms of the 'Buffy Paradigm,' and . . . think about many of the problems in the proposed solutions as part of the 'Buffy Syndrome.'"

Cordesman's point is that just as the characters in *Buffy* are always trying to create plans and models that jar with normative reality, we, too, live in a world where we never really face the level of uncertainty we have to cope with. His idea is that research speculation on scenarios, delivery methods, and lethality can be conducted until hell freezes over, but the best strategy is the kind of "expect the unexpected" approach of Buffy and her Scooby gang (a nod to the "pesky, meddling kids" in the old animated TV series, *Scooby-Doo*), at least until a much clearer picture of what kind of biotechnology and biological attacks actually materialize over the coming decades.

THE SCIENCE AND TECH OF THE BUFFYVERSE

When Buffy and her gang of Scoobies are faced with some new peril, they immediately kick into "science mode." Buffy gets guidance from her Watcher, Rupert Giles, a member of the Watchers' Council, whose job is to train and mentor the Slayers. Giles researches the supernatural phenomena Buffy encounters, offers educated speculation as to their provenance, proffers counsel on how to best them, and evolves a training program to help her remain fighting fit. The program makes clear that omitting the crucial step of the science mode often leads to failure. Just as scientists must understand the nature of a problem ahead of designing experiments to interrogate and confront its reality, Buffy and her Scooby gang know they must first understand the nature of whatever new evil comes to town in order to defeat it.

There are parallels in tech, too. Take the example of what we might call "Schrödinger's paper." On *Doctor Who*, a form of Schrödinger's paper, known as psychic paper, when shown to a person, could usually induce them to see whatever the user wished them to see printed on it. The Tenth Doctor explained that the paper "assigned authority based on the reader's perceptions." (Austrian physicist Erwin Schrödinger stated that if you placed a cat and something that could kill the cat, such as a radioactive vial, in a box and sealed it, you wouldn't know if the cat was alive or dead. The condition of the cat was essentially unknowable until you opened the box, so that until the box was opened, the cat was effectively both alive *and* dead.) In the Buffyverse, volumes in the library of Wolfram & Hart, "the devil's law firm," are blank until someone asks for a specific book. Only then do the empty pages fill with the relevant text. In the real world, electronic paper is an analogous tech, used for commercial signage in the marketplace.

In another narrative, Buffy uses a mirror to throw a witch's hex energy back onto the witch. This ploy is a similar technique to Alexander Graham Bell's photophone. This was a forerunner of fiber-optic communications, which transmitted sound on a beam of light to a mirror, making the mirror vibrate in response. The photophone then recaptured the vibrations thrown by the mirror and converted them back into sound.

There's also a digital form of demon, somewhat akin to a computer virus. Buffy's witch friend, Willow, scans some ancient text into a computer. But the text is a mystical work in which a demon is bound. When the text is scanned, the demon becomes binary, his essence broken into electron bits, then digitized into computer bytes and potentially unleashed on the internet. Buffy's Watcher, Giles, comes to the rescue, with the essential help of the school's computer science teacher, Jenny Calendar. They combine magic with data and defeat the demon by forming a virtual mystical circle in an online chat room that casts a "rebinding" hex.

THE ORIGINS OF HELL

The fusion of science and magic is a defining feature of the Buffyverse. And a perfect example of that fusion is the program's concept of hell. Joss Whedon has gone on record suggesting that *Buffy* was meant as a metaphor for how high school students can sometimes feel that school is like hell. Thus, Whedon dreamed up a literal hell in a fictional high school, complete with vampires and myriad monsters denoting humanity's inner demons.

Sunnydale High School's library was located directly above the hellmouth. It served as the base of operations for Buffy and her gang, and it was also where Giles practiced the art of Watcher. He brought in many old supernatural texts covering a number of religious and mythological visions of hell but, generally, in several ancient civilizations, hell is seen as a realm of punishment or suffering after death.

In ancient Mesopotamian mythology, particularly in the Sumerian epic poem *Gilgamesh*, which dates back to eighteenth century BCE, there is one of the earliest known references to an afterlife of punishment. *Gilgamesh* describes a dark underworld, the "House of Dust," where the dead are destined to dwell in eternal darkness and suffering. In ancient Egyptian lore, the idea of a realm of judgment and punishment after death was depicted in the *Book of the Dead*. The Egyptians believed in a complex afterlife journey, with the wicked facing censure and the righteous rewarded in the realm after death. The idea of hell also played a crucial role in Christian theology. In their *Bible*'s New Testament, Jesus of Nazareth talks about Gehenna, a place of punishment. Again, Christian visions of hell vary among the religion's denominations but, generally, hell involves eternal separation from God and abiding suffering as a sequel to sin.

THE SCIENCE AND MAGIC OF HELL

Bearing in mind this eternal separation from God, let's consider the fascinating scientific evolution of the ancient Greek ideas of heaven and hell. The Pythagorean philosopher, Philolaus, was the earliest known thinker to assign motion to the Earth. In his view, Earth was airborne. A contemporary of Socrates, Philolaus argued that numbers governed the Universe. He also did away with the idea of fixed direction in space and gave science one of the first non-geocentric models of the Universe through his creation of an object known as the central fire. His was a cozy cosmos, one in which you could put your feet up and warm them by the fire. No longer did the Earth stand central, massive, and immobile. At the hub of Philolaus's realm was "the hEarth of the Universe," the central fire, and around this core revolved nine bodies: the Earth, Moon, and Sun, the five planets, and the sphere of the fixed stars.

The central fire could never be seen, for the "civilized" and inhabited central region of the globe, the Greek world and its neighbors, was always facing away from the fire. Much like the far, dark side of the Moon was always facing away from the Earth. And beyond this cozy cosmos of the central fire was its main source of light, the outer fire. The outer fire bound the cosmos on all sides. A wall of fiery ether, it was an eternal spring of luminous energy. A little like Tolkien's "Eye of Sauron," Philolaus's Sun served merely as its portal or lens through which the outer light was drawn and dispersed. It was a fantastic notion. But perhaps no more incredible than what we think we know and believe today. Namely, the notion of a ball of burning gas careering across an endless sky.

Now, some of the ancient Greeks, particularly the Pythagoreans, realized that the daily revolution of the entire sky was merely an illusion, one caused by the Earth's own motion about the central fire, yet Philolaus did not take the next obvious step. He did not have the Earth rotate on its own axis. Admittedly, his cosmology is innovative enough for ancient times, but having already taken the revolutionary step of releasing the Earth, allowing it to be free and mobile in space, he did not see it spin.

It was not the cosmology of Philolaus that led to the scientific conception of hell, and that's because Philolaus lost the battle of the cosmologies to Aristotle. The Universe of the ancients had been a kind of cosmic oyster, clammed up in space and time. Pythagoreans like Philolaus tried to force it open, let a spherical Earth adrift, and introduce a cosmos of change. But Aristotle wound back the change, returned an immobile Earth to its center, and popped the genie back in the bottle.

UNRAVELING ARISTOTLE'S HELL

Aristotle's Universe had hard limits in time and space, and for two millennia his cosmology held sway. His vision was of a two-tier, geocentric cosmos. The Earth, mutable and corruptible, was placed at the center of a nested system of crystalline celestial spheres, from the sublunary to the sphere of the fixed stars. The sublunary sphere, essentially from the Moon to the Earth, was alone in being subject to the horrors of change, death, and decay. Beyond the Moon, all was immutable and perfect. The Earth was not just a physical center. It was also the center of motion, and everything in the cosmos moved with respect to this single center. Aristotle declared that if there was more than one world, more than just a single center, elements such as Earth and fire would have more than one natural place toward which to move, in his view a rational and natural contradiction, so Aristotle concluded that the Earth was unique.

Let's unravel hell, or at least Aristotle's version of it. After all, as Giles's library would attest, Aristotle's system stood the test of time. It was a synthesis of earlier Earth-centered cosmologies and had an immense influence on medieval religion and culture. His walled-in Universe encases nine concentric spheres, clear as crystal. The outermost sphere is that of God, the Prime Mover. It is God who spins the world from outside. The motion He imparts to the outermost sphere is transferred to each adjoining sphere in turn. Like a child's clockwork toy, the sky is reduced to a mechanical curiosity, God himself keeping the machinery in motion.

Philolaus filled the entire cosmos with a source of cosmic energy, the central fire. With God's removal to the outer limits, Aristotle now bestows upon the Earth the most lowly and humble place in the whole Universe, the central region. Only this innermost layer, the sublunary sphere, is privy to dreadful change, and the eternal separation from God is born.

Beyond the sphere of the Moon, all is serene. This supralunary region houses each of the planets in their God-induced motion around the Earth. It is the classic geocentric system of the ancestors, dancing to Plato's divine tune of regular motion in a perfect circle. To the ancient Greeks, this system gifted cosmic comfort to a frightened world. And later to medieval minds, the division of the cosmos into two, one part lowly, the other divine, one part flux, the other eternal, brought the illusion of strength in times of turmoil.

Caught in a crossfire of cosmologies, Aristotle's other gift is that of supreme compromise. His Universe is a union of worldviews, a meeting place of materialism and idealism, and a profound consolation to the minds of the meek, and so Aristotle's sublunary sphere is the domain of the materialists. It is the region of Aristotle's Universe that is the result of dynamic forces in constant flux. Here, matter is made up of various fusions of the four classical elements of Earth, water, air, and fire. Each has a natural place to be: Earth downward, fire upward, and air and water horizontal.

The four elements are agents of change. They forever transmute, the sublunary atmosphere replete with their fusions. Indeed, the make-up of the atmosphere is not pure air. Rather, its substance is a catalyst of change, which when set in train ignites to create meteors and comets. In short, the sublunary sphere is buzzing!

The rest of Aristotle's Universe is dormant; divine but essentially dull, for beyond the Moon, there is no change. The four terrestrial elements that provide the key to change in the sublunary sphere are absent, here in the supralunary domain. The fabric of the cosmos was quintessence, the fifth element. And the farther we fly out, from the Moon to Mercury and beyond, the purer the quintessence becomes. Until it meets its highest form in the sphere of Aristotle's God, the Prime Mover.

DIVINE COMEDY

Aristotle's cosmology of hell found its way into the Christian view of the Universe. Consider the *Divine Comedy*, a fourteenth-century epic poem by the Italian poet Dante Alighieri. Dante's famous work is a fictional account of the poet's epic journey through the contemporary Christian cosmos. The journey begins on the surface of the spherical Earth. Dante then descends into the Earth and through the nine circles of hell, all mirroring the nine spheres of heaven per the ancient cosmology of Aristotle, which still held sway in medieval times. Dante reaches his first destination, the most corrupt of all realms, the squalid center of the Universe, locus of the Devil and his legions.

The poet reemerges at the other side of the globe and wings his way through the aerial regions, passing through the terrestrial spheres of air and fire, then flying through each celestial sphere, speaking with the spirits that inhabit each of them. Ultimately, he approaches the Primum Mobile, the first moved thing, and the sphere that gives motion to the rest of the realm. Finally, Dante beholds this last sphere, the Empyrean, God's Throne.

The Universe of Dante's *Divine Comedy* is a mix of science and magic. The science is Aristotle's, but it was tailored to the medieval holy church and God himself. The only life encountered on this journey through heaven and hell is either terrestrial or divine. No other single soul stirs beyond the sphere of the Moon. No one lives in the Sun, and no Martians dwell on the red planet. The alien is nowhere. And for good reason. It is a Universe malformed by Christian symbolism. By his use of allegory, Dante implies that the medieval Universe could have no other structure than the Aristotelian.

In Dante's powerful poem, a masterpiece of world literature, Aristotle's Universe of spheres mirrors man's hope and fate. Both bodily and spiritually, like Sunnydale in *Buffy*, humanity sits midway. Our pivotal position in the hierarchical cosmic chain is halfway between the inert clay of the Earth's core and the divine spirit of the Empyrean. The rest of the cosmos is made of either matter or spirit. But uniquely, humanity is made of both, body and soul.

Humanity's place is also transitional. In Dante, as in *Buffy*, we humans live on Earth, in filth and insecurity, close to hell. But we are at all times and in all places under the all-seeing eye of God, and with full knowledge of the heavenly escape above. Humanity's duality and place in the great scheme of things helped impose the dramatic choice that faced all Christians in the medieval world. Whether to follow their base and human nature down to its natural place in hell, or to engage with the spirit, and follow the soul up through the celestial spheres to God.

And what a wonderful history of hell upon which to base the story of a female superhero!

CHAPTER 3

HUMAN AND MONSTER: WHAT SUPER-FEMALES ARE IN ALIEN?

(. . . in which we meet the hero of a sci-fi/horror classic and her monstrous nemesis)

Everyone's gender in the script was deliberately left up in the air. I figured that the gender of each character would be determined at the time they were cast, and I wrote that into the first script, it's right there on the last page. Originally, John Travolta was considered to play Ripley, then out of nowhere "they came up with this Sigourney Weaver gal. They actually did a screen test with her, and everybody was favorably impressed." I don't see it as that revolutionary to cast a female as the lead in an action picture. It didn't boggle me then, and it doesn't boggle me now. My conception from scratch was that this would be a coed crew. I thought there was no reason you had to adhere to the convention of the all-male crew anymore. Plus, it was in

1976 that I was writing the thing, and it seemed like an obvious thing to do. I mean, *Star Trek* had women on for years.

—Dan O'Bannon, screenplay writer for *Alien*,
quoted in *Lip Magazine* (2015)

ALIEN

Ridley Scott's 1979 movie, *Alien*, is widely thought to be one of the greatest and most influential science fiction and horror films of all time, joining other movie classics, such as *Citizen Kane, 2001: A Space Odyssey*, and Alfred Hitchcock's 1958 masterpiece, *Vertigo*. By the year 2002, Scott's picture was considered "culturally, historically, or aesthetically significant" by the Library of Congress, and was duly chosen for preservation in the National Film Registry. In 2008, *Alien* was ranked by the American Film Institute as the seventh best film in the sci-fi genre, and as the thirty-third greatest movie of all time by British film magazine *Empire*. The success of *Alien* spawned its own media franchise, including further films, novels, comic books, video games, toys, and collectibles.

Alien tracks the tale of a small but doomed commercial freighter crew. They are diverted from their delivery and ordered by their corporate paymasters to check out a mysterious radio beacon. Their ship, *Nostromo*, was named by the scriptwriters after Joseph Conrad's novel of the same name. The ship's 182 model 2.1 terabyte artificially intelligent computer mainframe, MU-TH-UR 6000, was known to the crew simply as "Mother." Mother's ship sets down on planetoid LV-426. Like divers at the bottom of a dark sea, three of the crew, Captain Dallas, Executive Officer Kane, and Navigational Officer Lambert, make their way to the source of the radio signal, with wind and dust driving down in dark sheets. They discover an enormous but derelict alien craft. When the party of three enter the ship, they find its pilot is dead, but, in a high-ceilinged chamber whose walls are covered with shadowy lattices, they unearth a whole set of leathery ovoid shapes.

FACEHUGGER

While peering into and touching one of the egg-like structures, Executive Officer Kane is attacked by one of the creatures from inside the egg. In one of the most famous scenes in all of cinematic history, and in a shocking kind of jack-in-the-box manner, the creature lunges onto Kane's head, penetrates his helmet, and the so-called "facehugger" attaches itself to his face. (As the original *Alien* script put it "with shocking violence, a small creature smashes outward. Fixes itself to his mask. Sizzling sound. The creature melts through the mask. Attaches itself to Kane's face. Kane tears at the thing with his hands. His mouth forced open. He falls backward.")

Dallas and Lambert carry the unconscious Kane back to the *Nostromo* with, as the script tells us, "the life form still wrapped motionless around his face." Science Officer Ash curiously breaks quarantine by allowing them back on board. The crew examines Kane who is being kept alive in symbiosis with the creature. Eventually, the facehugger falls off of Kane and he seems to be fine.

CHESTBURSTER

With *Nostromo* a full ten months from Earth, and traveling at light speed, the stars approaching the ship appear blue while receding stars run to amber due to the craft's velocity. Kane finally comes to in the infirmary, feeling like "somebody's been beating me with a stick for about six years. God, I'm hungry." Soon the entire crew is seated in the mess, hungrily swallowing huge portions of artificial food. Then, Kane grimaces. His voice strains. "I'm getting cramps." The whole crew stares at him in alarm. Kane makes a beastly noise. He clutches the edge of the table with his hands, knuckles whitening. His face screws into a mask of agony as he falls back into his chair. A red stain appears on his chest, then smears into blossoms of blood as the fabric of his shirt is ripped apart. A fist-sized bestial head with metallic teeth pushes out. The crew yells in panic, and leaps back from the table. The creature's head lunges forward and comes spurting out of Kane's chest, trailing a serpentine body, splattering fluid and blood in its wake. As the "chestburster" disappears from sight, Kane lies slumped in his chair, a huge hole in his chest, stone dead.

XENOMORPH

The rest of the film entails the remaining crew hunting the creature down. As the alien, also known as the xenomorph, slowly kills the crew one by one, ultimately Ripley is the last human standing. During an initial draft of the script, there was a thinly sketched all-male gender characteristic of the crew. The intention was to directly challenge male audience members with the movie's messages of terror at the castration of the male crew by the female xenomorph (about which, more later).

Enter Alan Ladd Jr., film industry executive and producer at Fox Studios. It was Ladd who reportedly suggested the change that would forever leave its mark on the genre of science fiction: What if Ripley was a woman? What if the hero of the movie was female, like the xenomorph? Ladd's educated guess was that movie audiences would engage with a woman in peril. And so, out of the moviemaking process, one of cinema's most powerful female characters was created.

RIPLEY, SUPERWOMAN

Alien became one of the first action movies in which a woman plays the representative hero. The character of Sigourney Weaver's Ripley combines the survivor of slasher films with the heroic astronaut of traditional sci-fi. The monster in the form of the xenomorph is also female. An unrelenting force of nature, the xenomorph is not only a killing machine, but also a relentless reproductive machine. In some ways, the whole plot of *Alien* is built around the xenomorph's mothering and reproductive functions. In this sense we can see that her treatment of the male crew member Kane—penetrated, impregnated, and made to give birth—is a form of rape, from the initial attack of the *facehugger* to the explosive birth of the *chestburster*. Author David McIntee is one writer who holds to this idea. In his book, *Beautiful Monsters*, McIntee says "*Alien* is a rape movie with male victims. It shows the consequences of that rape: the pregnancy and birth. It is a film that plays, very deliberately, with male fears of female reproduction." This plot feature, which revolves around the idea that men can be impregnated and subsequently forced to give birth, is one of the most radical features of *Alien*. It highlights the fact that males have become the victims, and women the heroes, which is further articulated with the story of Ripley that follows. At the start of the movie, Ripley had been just one of seven faces, but by the end of the film she's the sole survivor.

Ripley's final battle with the xenomorph in *Alien* is key to her character as a superwoman. On hearing the deaths of the last two remaining crew members over the intercom, Ripley tries to prevent Mother from initiating the ship's self-destruct protocol. Ripley fails. While screaming that Mother is a "bitch," Ripley then gets into the shuttle *Narcissus* (another Joseph Conrad reference) and prepares for her escape as *Nostromo* self-destructs.

BEAUTY AND THE BEAST

Ripley prepares for hyper-sleep, but the xenomorph returns, so Ripley must face the monster for one final affray. For the first time in the movie, Ripley is openly sexualized on screen. She is seen in only underwear and a vest, as we become aware of her being sexualized and objectified as a woman. Meanwhile, one of the limbs of the xenomorph falls out of a crevice within the cramped escape craft. We see the terror in Ripley's face as the xenomorph starts to make its move. Ripley backs off, ever so slowly, and creeps into a nearby closet. We see her first recognize a helmet, then a space suit, as little by little she climbs into the suit. The movie cuts to the xenomorph's monstrous head, as its inner jaw extends toward the nearly naked Ripley. (Script: A clear glass panel in the door. The Alien puts its head up to the window. Peers in at Ripley. Their faces only two inches apart. The Alien looking at Ripley almost in curiosity.) Beauty and the beast. In the film's last few minutes, Ripley is vulnerably human, altogether terrified, just as we, the audience, would be.

Ripley finally secures the helmet and seals the space suit. With unbelievable bravery, she reaches for a grappling gun and inches toward the craft's control seat. She gradually buckles herself in, keeping a close eye on the peril at large. Once seated at the control panel, Ripley is able to vent the craft's gasses in an attempt to push the xenomorph from its hiding place. Each venting of gas brings the monster closer until we witness a camera close-up of Ripley's head as the xenomorph's inner jaw comes into view. Just as the xenomorph's monstrous mouth is about to munch Ripley's helmet, she blows the hatch, almost sucking the monster into space.

The rush of cabin air forces her control seat to swivel, so now she sits face to face with the monster, which is still clinging to the craft's interior. Ripley grabs the grappling gun and blasts the monster in the chest, sending the beast into space. But on releasing the gun, she finds the weapon is trapped in the closing door, allowing the xenomorph to claw onto the side of the craft. As the alien uses its barbed tail to gain hold of the craft and crawl into one of its engines, Ripley punches the control panel and barbecues the beast alive, finally making the spacecraft safe.

> Script: The Creature struggling. Jet exhaust located at the rear of the craft. The engines belch flame for a few seconds. Then shut off. Incinerating, the Alien tumbles slowly away into space. The burned mass of the Alien drifts slowly away. Writhing, smoking. Tumbling into the distance. Pieces dropping off. The shape bloats, then bursts. Spray of particles in all directions. Then smoldering fragments dwindle into infinity.

With the craft now repressurized, Ripley is finally safely seated in the control chair. Calm and composed, almost cheerful. She speaks into the ship's recorder:

> Final report, the commercial starship *Nostromo*. Third officer reporting. The other members of the crew, Kane, Lambert, Parker, Brett, Ash, and Captain Dallas are dead. Cargo and ship destroyed. I should reach the frontier in about six weeks. With a little luck the network will pick me up. This is Ripley, last survivor of the *Nostromo*, signing off.

She switches off the recorder. Stares into space. The shuttle-craft *Narcissus* sails into the distance with Ripley finally in hyper-sleep.

THE BECHDEL TEST

Any reckoning of Ellen Ripley as a superwoman of cinema must include the fact that the movie *Alien* was one of the inspirations for the so-called Bechdel Test. The Test became famous on the internet as a means to test whether a movie can be seen as feminist. The rules are straightforward enough: If a movie, (a) has two or more female characters, (b) who actually talk to one another, (c) about something other than a mere man, or have a conversation that actually moves the plot forward, the movie passes the test.

The rules of the Bechdel Test first appeared in 1985 in Alison Bechdel's comic strip, *The Rule*. The strip features two women discussing movies and one woman says that she only goes to a movie if it satisfies the requirements listed above from (a) to (c), which meant that the last movie she was qualified to watch was *Alien*, which, of course, had been released a full six years before the comic strip.

The Bechdel Test has become influential in the discussions of feminist cinema. Sweden has instituted the Test as an integral part of their national ratings system, and there are plenty of sites online devoted to listing which movies pass the Test and which ones don't. There are, naturally, other means for analyzing whether a movie is feminist or not. (For example, the "Sexy Lamp Test" suggests that if a female character can be replaced by nothing more than a sexy lamp, the movie has gender bias and weak female characters.) But the fact that the Bechdel Test singled out *Alien* as passing the initial test is important to its place in the history of feminist cinema.

FEMALE HUMAN VERSUS FEMALE ALIEN

The character of Ellen Ripley has another special place in the history of sci-fi cinema. Her struggle is one of the greatest ever examples of the "human versus alien" motif. Let's dig a little deeper on that special place in sci-fi history. Today, Earth is an "alien" planet. It has been so since the Scientific Revolution, since astronomers found that Earth no longer sat at the center of the cosmos. That discovery not only made Earths of the planets, it also effectively brought the alien down to Earth. How? Well, the Universe of our ancestors had been small, static, and Earth-centered. You could say it had the stamp of humanity about it. In ancient cultures, the constellations were given the names of Earthly myths and legends, and the magnificence of the cosmos gave evidence of God's glory.

But the new Universe unveiled by science was inhuman. The farther out the telescopes probed, the darker and more alien it became. "The history of astronomy," suggested British novelist Martin Amis, "is a history of increasing humiliation. First the geocentric Universe, then the heliocentric Universe. Then the eccentric Universe—the one we're living in. Every century we get smaller. Kant figured it all out, sitting in his armchair . . . The principle of terrestrial mediocrity."

American astronomer and science fiction writer, Carl Sagan, had gone further. Sagan saw that humans had suffered a series of "Great Demotions" in the last five centuries. First there was Earth. It was not at the center of the Universe, nor was it the only object of its kind, made of a unique material only to be found on terra firma. Next came the Sun. Not at the center of the Universe, not the only star with planets, nor eternal.

There were more stars in this new Universe than grains of sand on all of Earth's beaches. The Milky Way Galaxy too proved neither at the center of the cosmos, nor the only galaxy within it. Two trillion other galaxies have since been discovered; adrift in an expanding Universe so immense that light from its outer limits takes longer than twice the age of the Earth to reach terrestrial telescopes . . . and there may be other Universes. The final demotion, Sagan suggested, would be the discovery of another biological intelligence in the Universe. And that's exactly what Ripley finds.

Consider how German astronomer and mathematician Johannes Kepler had first reacted in the early 1600s when he heard of Galileo's discovery of other worlds in the form of the (previously unknown) moons of Jupiter and the existence of Earth-like craters and mountains on the Moon. Kepler felt an immense sense of wonder. And he also felt a huge sense of *estrangement* from this new reality. Estrangement in the sense of a state of imperfect knowledge. The result of coming to understand what is just within our mental horizons. Kepler felt that Galileo's discoveries meant that creatures may well be dwelling on these other worlds. He felt it was a vital new piece of evidence in the debate on the existence of alien life.

And it was Kepler who motivated H. G. Wells to write, nearly three hundred years later, "But who shall dwell in these Worlds if they be inhabited? Are We, or They, Lords of the World? And how are all things made for Man?" (And, we might add, is Ripley or the xenomorph "Lord of the World?") The same sense of wonder at the discoveries made by the likes of Galileo and Kepler is also common to science fiction.

Science fiction effectively began with the early discoveries of the Scientific Revolution. It marks the paradigm shift of the old Universe into the new. The old and cozy geocentric cosmos was about us humans. The new Universe of Kepler and Galileo was decentralized, inhuman, infinite, and alien. Historically, then, science fiction is a response to the cultural shock created by the discovery of humanity's marginal position in a Universe fundamentally inhospitable to humans. Science fiction is our attempt to make human sense of the new Universe.

How does science fiction like *Alien* work? By conveying the taste, the feel, and the human meaning of discovering our marginal position in an alien Universe. Conveying the cultural shock of the new discoveries of science. In a way, sci-fi is like old-school religion. Both are concerned with the relationship between the human and the nonhuman. In the case of religion, its interest is the relationship between the human and the divine. With sci-fi, its chief concern is a response to the demotions of life in the Universe and is a kind of displacement of religion. In this way, science fiction may be viewed as the "soul" of science, with its focus on the human-nonhuman opposition.

SPACE, TIME, MACHINE, AND MONSTER

The subgenre of superwomen is placed firmly within the genre of science fiction. And sci-fi presents us with an infinity of future visions. A dazzling diversity of contrasting elements: aliens and time machines, spaceships and cyborgs, utopias and dystopias, androids and alternative histories. But, on a more thoughtful level, we can identify four conceptual themes: *space, time, machine,* and *monster.* Each of these themes is a way of exploring the relationship between the human and the nonhuman.

Within the science fiction of the *space* theme, space can be represented by the alien, an animated version of nature, where the likes of the xenomorph are agents of the void, in a similar way to how the shark in *Jaws* is a deadly creature of the deep.

The science fiction of the *time* theme portrays a flux in the human condition fashioned by some process that is revealed in time. For example, in *Action Comics #255*, a comic from the 1950s, Supergirl traveled through time to the twenty-first century to experience what life is like in the future (needless to say, the depiction of twenty-first century life is woefully inaccurate).

The sci-fi stories within the *machine* theme include tales about robots, computers, and artificial intelligence. One of the most influential early films featuring a robot was Fritz Lang's 1927 movie, *Metropolis.* It included *maschinenmensch* (literally "machine-human" in German), a robot in (female) human guise. Popularly known as "false Maria," maschinenmensch was one of the first fictional robots ever depicted in cinema, and consequently popularized the idea worldwide. Dystopian tales are part of the man-machine theme; it is the *social machine* in which the human confronts the nonhuman.

Finally, sci-fi stories featuring the nonhuman in the form of *monster* are usually situated within humanity itself. This is the category in which superwomen and superheroes generally lie. In these tales, there is often an agency of change that leads to the change of human into nonhuman. For instance, the agency of change for Supergirl is simply her transfer from Krypton to Earth, which leads to a multitude of superhuman abilities derived from the rays of our yellow sun. And a transfusion of gamma-irradiated blood from her cousin Bruce Banner, a.k.a. the Hulk, gift Jennifer Walters, the She-Hulk, *her* superhuman powers.

HUMAN VERSUS ALIEN

Ellen Ripley's superhuman battle with the xenomorph should be seen in this sci-fi context of exploring the relationship between the human and the nonhuman, with its *space, time, machine,* and *monster* themes. Figure 1 here may help chart Ellen Ripley's confrontation. At times, as with 2019's Chinese movie blockbuster, *The Wandering Earth,* science and the human are pitched against nature and the nonhuman. In such cases, the nonhuman comes in the form of a force of nature, which shatters the stability of the human world. In *The Wandering Earth,* it's the year 2058, and an anomalously expanding red giant Sun threatens to engulf the Earth within one hundred years. In dystopias, such as the 1999 movie, *The Matrix,* nature and the human are united in collision against science and the nonhuman machine world. As *The Matrix* suggests, sci-fi can sometimes characterize science as unnatural and nonhuman. The natural and organic human home-world of Zion collides with the mechanical and scientific alternate world of The Matrix itself. According to this convention, utopias are imagined societies that are more fully human than the present.

More often, though, science features on both sides of the human-nonhuman conflict. In H. G. Wells's *The War of the Worlds,* for example, science is part of the nonhuman element symbolized by the invading Martians. Like the xenomorph, they are agents of the void. They also embody science with their vast, cool, and unsympathetic

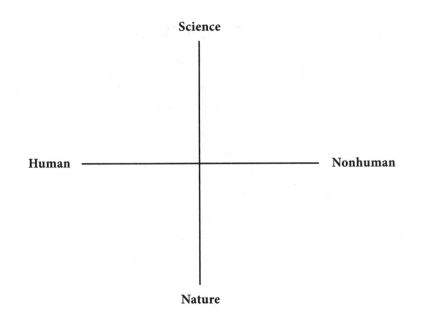

Figure 1. The worldview of science fiction, after Professor Mark Rose.

intellects. Later, however, these alien invaders fall victim to microbes, a fate which pits the science and understanding of unsolicited natural selection on the side of the invaded humans. We can see parallels between Wells's *The War of the Worlds* and Ridley Scott's *Alien*. The xenomorph is a biomechanical force of nature, and has since become a pop-culture touchstone that has already gone down in cinema history as iconic. The xenomorphs are deadly, highly intelligent, raptorial monsters that have but one purpose in mind: to sustain their Hive. They even look like metallic machines. They embody science in the sense that we understand how unsolicited natural selection has molded this monster into the perfect predator. But Ripley also uses science and cunning to defeat the xenomorph: venting her spacecraft's gasses in an attempt to push the xenomorph from its hiding place, blowing the hatch in an attempt to suck the creature out into space, and punching the control panel to barbecue the beast alive, finally making the spacecraft safe.

THE HIERARCHY OF FOREIGNNESS

Ripley's is no small victory. And to get a good idea of just how big a victory it was, consider the idea of the "hierarchy of foreignness." The hierarchy comes from the 1986 Orson Scott Card novel, *Speaker for the Dead*, and is a classification system of humans and other species.

The hierarchy is arranged in four tiers, from least alien to most. The first tier is an *utlänning*, meaning "foreigner" in Swedish, and is defined as a stranger recognized as human from the same planet as a subject but of a different nation or city. For example, an American would be an *utlänning* to a native of India. The second tier is a *främling*, meaning "stranger" in Swedish, and is defined as a stranger recognized as human but from a different planet than a subject. For example, a human from a colony on Mars would be an *främling* to a native of Earth. The third tier is a *ramen,* meaning "the framework" in Swedish, and is defined as a stranger recognized as human but of another sentient species entirely. This one is a little trickier to define in simple terms, but maybe we could say that Neanderthals, the extinct subspecies of archaic humans also known as *Homo neanderthalensis*, would be *ramen* to a humans, *Homo sapiens*. The term *ramen* was only ever used to refer to the entire species as a whole, rather than an individual member. Finally, the fourth tier is v*arelse*, meaning "creature" in Swedish, and is defined as truly alien—they may, or may not, be intelligent sentient beings. But they are so foreign that no meaningful communication is possible with the subject.

JAWS IN SPACE

"Jaws in space" was the original studio pitch by *Alien* screenwriter, Dan O'Bannon, to sell his soon-to-be science fiction classic. The hierarchy of foreignness above confirms O'Bannon's pitch. Consider some telltale aspects of Ripley's "meeting" with the xenomorph. Not only was the xenomorph "alien," but it was *truly* alien; it was *varelse*. Consider also the bestial behavior and animalistic characteristics of the xenomorph that Ripley contends with. The xenomorph is a highly aggressive creature that attacks without provocation. Its primary mode of attack is to use its sharp claws and teeth to tear apart its prey. The creature is incredibly strong and fast, which allows it to overpower its victims super quickly. It has heightened animal-like senses, including excellent hearing and sense of smell, which it uses to track and hunt its prey; the prey being all the crew members of the *Nostramo*. The xenomorph has a reptilian appearance with a long, flexible tail, sharp claws, and a mouth that can extend and retract. Its appearance is reminiscent of a creature from prehistoric times, which adds to its bestial qualities. So, the xenomorph is a *varelse* alien that exhibits many traits commonly associated with bestial creatures. Indeed, it could quite easily be argued that the character of Ripley had to deal with a type of monster not previously seen on screen.

Ellen Ripley, superwoman, is frequently featured in lists of the best characters in film history. For example, in their 2008 list of the one hundred greatest heroes and villains in American film history, the *American Film Institute* ranked Ripley as the eighth best hero, the second-highest ranked female character after Clarice Starling. Ripley was ranked fifth in *Entertainment Weekly*'s 2009 poll of the twenty all-time coolest heroes in pop culture, and Ripley was referred to as "one of the first female movie characters who isn't defined by the men around her, or by her relationship to them." Ripley was also ranked ninth in *Empire* magazine's compilation of the one hundred greatest movie characters in 2008, and fifth in 2015; the highest ranked female in both polls.

The accolades just keep on coming. Ripley was ranked eighth in *Premiere* magazine's poll of the one hundred greatest movie characters of all time. *Premiere* described her defining moment as her "nervy refusal to open the ship's hatch so that Kane can be admitted—with a thing attached to his face." She was the third-highest ranked female on *Premiere*'s poll, after Annie Hall and Scarlett O'Hara. In 2009, MTV selected Ripley as the second greatest movie badass of all time, the only woman with Sarah Connor, who was ranked sixth. In 2011, *Total Sci-Fi* ranked Ripley first on their list of the twenty-five women who shook sci-fi, calling her "one of the most iconic characters in cinema history" and "one of the most critically analyzed characters in the history of cinema." Finally, in 2011, *Total Film* ranked Ripley the best female character of *any* movie. Superwoman.

CHAPTER 4

HOW HAS WONDER WOMAN CHANGED WITH THE TIMES?

*(. . . in which we look at the social, psychological,
and political science of Diana Prince)*

William Moulton Marston, a psychologist already famous
for inventing the polygraph, struck upon an idea for a new
kind of superhero, one who would triumph not with fists
or firepower, but with love. "Fine," said [wife and fellow
psychologist] Elizabeth. "But make her a woman." From her
lips to his drawing board. Wonder Woman made her début in
December 1941 in *All Star Comics*, a bimonthly with strips by
different artists . . . The first episode features Diana rescuing
US Army Intelligence Officer Steve Trevor, whose plane has
crashed on uncharted Paradise Island. Aphrodite and Athena,
the ruling goddesses of the Amazons, command that the
"strongest and wisest" she-warrior return Steve to America,
and there remain to defend the "last citadel of democracy,
and of equal rights for women." Diana wins the honor, besting
her sisters in an Amazon Olympics, and so begins her close to

sixty years and counting of fighting for "liberty and freedom for all womankind."

—Marguerite Lamb, "Who Was Wonder Woman 1?"
(*Bostonia* magazine, 2001)

FAVORITE SUPERHEROES

Comic books are a global phenomenon, and the cinematic adventures of comic book characters are a billion-dollar industry. To date, the Marvel Cinematic Universe movies have grossed almost $30 billion, Spiderman over $10 billion, the Avengers almost $8 billion, Batman roughly $7 billion, the DC Extended Universe around $6.4 billion, and X-Men just over $6 billion. You get the picture. Comic books and their superheroes mean big bucks, so it's hardly surprising that the ever-burgeoning fan base is always on the lookout for the latest and greatest "caped crusader."

Of all the hundreds of superheroes in film and fiction, who are the current favorites? *Wise-voter* recently conducted a poll to find America's favorite comic book superhero. The results were very telling (see Table 1 below.) Of all the states, twenty-five voted Spiderman as their favorite, which means that Marvel currently has a comfortable lead in the age-old battle between Marvel and DC. Batman came a distant second to Spiderman, with Batman polling top in ten states, followed by Superman, who came in third by topping the poll in four states. In total, Marvel came out king, taking thirty-two states.

State	Superhero
Alabama	Wolverine
Alaska	Captain America
Arizona	Spiderman

Arkansas	Superman
California	Spiderman
Colorado	Spiderman
Connecticut	Spiderman
Delaware	Captain Marvel
Florida	Spiderman
Georgia	Spiderman
Hawaii	Batman
Idaho	Superman
Illinois	Spiderman
Indiana	Spiderman
Iowa	Aquaman
Kansas	Superman
Kentucky	Spiderman
Louisiana	Spiderman
Maine	Wonder Woman
Maryland	Spiderman
Massachusetts	Batman
Michigan	Batman
Minnesota	Batman
Mississippi	Spiderman
Missouri	Spiderman
Montana	Wolverine
Nebraska	Spiderman
Nevada	Batman
New Hampshire	Spiderman
New Jersey	Spiderman
New Mexico	Iron Man
New York	Batman
North Carolina	Spiderman
North Dakota	Aquaman
Ohio	Spiderman
Oklahoma	Batman

(Continued on next page)

Oregon	Black Panther
Pennsylvania	Batman
Rhode Island	Captain Marvel
South Carolina	Spiderman
South Dakota	Spiderman
Tennessee	Spiderman
Texas	Spiderman
Utah	Spiderman
Vermont	Wonder Woman
Virginia	Spiderman
Washington	Spiderman
West Virginia	Batman
Wisconsin	Batman
Wyoming	Superman

Table 1. Favorite comic book superhero by state

While the superhero industry has never been more profitable, female superheroes certainly seem to be hugely underrepresented in film, fiction, and fandom. Given that Captain Marvel has, at different times, been both male and female, the only superwoman who comes out on top in any state in the *Wise-voter* poll is Wonder Woman, who won the votes in Maine and Vermont.

WONDER WOMAN

The question of why Maine and Vermont bucked the national trend in the US is not our concern here, though it remains an interesting question. A more pertinent question for this book is, why Wonder Woman? Well, historically, Wonder Woman is usually the first superwoman people think of, whether or not they have ever read a comic book or seen superheroes on television or in the cinema. Wonder Woman is the most popular superwoman of all time. Batman and Superman aside, no other comic book character has lasted as long. Generations of children have carried their food to school in Wonder Woman lunch boxes. Just as Superman is synonymous with all things "super," Wonder Woman evokes the idea of a woman who can do it all, and do it well. Indeed, she is often considered something of a feminist icon, a totem of female power, symbolizing equality with her male counterparts.

From the get-go, what made Wonder Woman different? Sure, she exhibited strength, bravery, and resolute decision-making in her heroism, fighting for truth and justice and protecting the innocent. But Batman, Spiderman, and Superman did that too. Still do. So that's not what makes Wonder Woman different. What made Wonder Woman unique when she first hit comic books back in 1941 was that she channeled her power through her female form. This made her special. She was set against the grain of long-standing male comic book characters and was considered potentially threatening, even though she was written as an amelioratory character, approaching all situations with an open mind, an open heart, and an open hand. Indeed, ever since her inception, in an evolving world in which women have been treated unequally in law and custom for way too long, Wonder Woman has shown that having brains, brawn, and the caliber to command are not merely male traits. No, they are *human* traits. Traits that can be performed by anyone. Anyone can be heroic.

On the other hand, let's think a little more about the image of Wonder Woman. What does she seem to portray? If we take her image and status literally, the example of Wonder Woman seems to suggest that superwomen are white and of royal origin, are necessarily near naked and drop-dead gorgeous, and certainly straight and able-bodied. It's too easy to interpret her persona as representing a host of privileges that the vast majority of women simply don't possess. Invented, imagined, and pictured mostly by men, for a long time Wonder Woman was frequently the sole female character in a particular tale, set adrift in a sea of men. Habitually sexualized, her rendering has often failed to focus on the idea that women in general are equal to men. Rather, and this of course is the general trouble with the exceptionalist nature of superheroes, it's just this specific bathing-suited beauty of a white woman who can overpower men.

The backstories of male superheroes have rarely, if ever, historically inspired social-science discussions about the place of men in society, but Wonder Woman has oftentimes underlined rather traditional roles about women, while at other times creating a debate for more fluxional gender potentials. On occasion, both have occurred simultaneously. So, let's take a look at the evolution of Wonder Woman from her progressive beginnings in the comics of the 1940s, her domestication in the 1950s and 1960s, and her migration into other media forms from the 1970s on.

THE ORIGIN OF WONDER WOMAN

October, 1940. In an interview with *Family Circle* magazine, American psychologist William Moulton Marston speaks prophetically about the potentiality of comic books. Marston, also known by the *nom de plume*, Charles Moulton, had invented an early prototype of the lie detector, along with his wife Elizabeth Holloway (the two had worked on the development of a systolic blood pressure measurement used to detect deception; the predecessor to the polygraph). The *Family Circle* article caught the attention of comic book publisher, Maxwell Charles Gaines, who hired Marston as an educational consultant for two of the companies that would merge to form DC Comics.

Marston had the bright idea of creating a new superhero. Naturally, it was his wife, Elizabeth, who convinced him that the superhero should be a woman. After all, the couple's experience with polygraphs had convinced them that women were more honest than men in specific situations and could work more efficiently, so the idea of Wonder Woman was developed, whom Marston considered to reflect his era's more unconventional and liberated women.

Wonder Woman was designed to be a parable for the ideal love leader, a philosophy of leadership based on a kind of Biblical love, the highest form of love, "the love of God for man and of man for God." Marston held that Wonder Woman would be the kind of woman who should run society. As Marston once put it, "frankly, Wonder Woman is psychological propaganda for the new type of woman who, I believe, should rule the world." Marston also drew creative inspiration from the bracelets worn by Olive Byrne, who lived with the couple in a polyamorous relationship.

Marston explained his creation more fully in a 1943 issue of the quarterly literary magazine, *The American Scholar*:

Not even girls want to be girls so long as our feminine archetype lacks force, strength, and power. Not wanting to be girls, they don't want to be tender, submissive, peace-loving as good women are. Women's strong qualities have become despised because of their weakness. The obvious remedy is to create a feminine character with all the strength of Superman plus all the allure of a good and beautiful woman.

THE HISTORY OF WONDER WOMAN

By the summer of 1942, in the middle of WWII, a press release from the New York offices of *All-American Comics* materialized at magazines, newspapers, and radio stations all over the US. "Noted Psychologist Revealed as Author of Bestselling Wonder Woman," it declared. The truth about Marston as the creator of Wonder Woman was finally out.

In the very first issue, a fictional newspaper editor by the name of Brown is desperate to discover Wonder Woman's secret history, so he sends out a scoop of journalists and reporters to chase her down. Needless to say, Wonder Woman not only evades them, but she also disguises herself as a nurse and visits Brown in hospice, recuperating from being driven half-mad by Wonder Woman's evasion. She presents Brown with a scroll. "This parchment seems to be the history of that girl you call 'Wonder Woman,'" she explains. "A strange, veiled woman left it with me."

Seemingly instantly recovered (this is a comic book, after all), Brown bounds out of bed and dashes back to work. "Stop the presses! I've got the history of Wonder Woman," he bellows. Curiously, the so-called secret history alluded to in the first ever Wonder Woman comic strip is as dramatic as the private life and papers of creator Marston.

The same Max Gaines who hired Marston as an educational consultant had been an elementary school principal before he founded *All-American Comics* and more or less invented comic books in 1933. Superman first sped faster than a bullet in 1938. Batman began prowling the streets of Gotham in 1939. The tales went viral among the country's kids, but some parents became concerned by the level of violence, even sexual violence, which comic books seemed to celebrate. (There was, after all, a very bloody and violent war going on in old Europe.) In this context, the *Chicago Daily News* in 1940 called comics a "national disgrace;" "ten million copies of these sex-horror serials are sold every month!" The paper's literary editor called for teachers and parents to ban such publications "unless we want a coming generation even more ferocious than the present one."

To defend himself against such criticism, Gaines hired William Marston as a consultant after first coming across him when *Family Circle* magazine sent a female staff writer to visit Marston at his home in Rye, New York, to ask his expert opinion on comic books. Gaines was proud of his acquisition. "Dr. Marston has long been an advocate of the right type of comic magazines," he said. After all, Marston *was* an outstanding candidate. He held three degrees from Harvard, including a doctorate in psychology. He led an "experimental life," associated with the experiential perspective of having been a lawyer, a scientist, and a professor. And he was credited with uncovering other people's secrets by inventing the lie detector test. He'd also act as a consulting psychologist for Universal Pictures and pen a novel, ten screenplays, and dozens of magazine articles.

William Marston and Wonder Woman were crucial to the creation of what was to eventually become *DC Comics*, the DC part of the title being short for *Detective Comics*, the publication in which Batman began. And Gaines resolved to counter his critics by setting up an advisory board, also appointing Marston to sit on it. DC soon stamped comic books in which Superman and Batman appeared with a logo, a badge of quality, which read "A DC Publication." In Marston's opinion, "the comics' worst offense was their blood-curdling masculinity," and the perfect way to silence the critics was to fashion a female superhero. Gaines's reply tossed the ball back into Marston's court. "Well, Doc, I picked Superman after every syndicate in America turned it down. I'll take a chance on your Wonder Woman! But you'll have to write the strip yourself."

THE AMAZONS

Early in 1941, Marston submitted a draft of his first Wonder Woman tale. She was introduced as "lovely as Aphrodite, as wise as Athena, with the speed of Mercury and the strength of Hercules" (Marston and Harry G. Peter 1941–42, *All Star Comics* #8). There was also an outlining of the subtext and backstory of Wonder Woman's Amazonian origins in ancient Greece, a place where men had kept women in chains, until they broke free and liberated themselves. "The new women thus freed and strengthened by supporting themselves (on Paradise Island) developed enormous physical and mental power." Marston claimed his comic would chronicle "a great movement now underway—the growth in the `power of women."

Who were the Amazons? Either wholly or partly mythical, the Amazons appear as early as Greek literature itself appears, so they are a Greco-Roman phenomenon. Over a period of more than a thousand years, starting with Homer, the Greek poet who lived in the eighth century BC, there were legends of a group of female hunters and warriors, as skillful and as brave as men in physical strength, agility, archery, horse-riding skills, and the science of combat. They were said to have lived around the Black Sea, starting off south of the sea and later moving north. There's also a tradition that the Amazons hailed from Libya, but the main point is that this culture was geographically marginal to the Mediterranean-based maritime empire of the ancient Greeks. No one knows where the word "Amazon" comes from; it isn't Greek, but there are a few options, one of which is an ancient Iranian word that means "warrior."

Legend had it that the Amazons were on the margins of both the Greek and Roman empires. They were culturally liminal. One source of their legend is fifth-century BC Greek writer, Herodotus, sometimes known as the father of history (a title conferred on him by the ancient Roman orator, Cicero). In his writing about the history of the Greco-Persian Wars, which began in 499 BC and lasted until 449 BC, Herodotus described the rise of the Persian Empire. In his account of one of the (unsuccessful) military campaigns of the Persian ruler, Darius the Great (522-486 BC), Herodotus talks about the Amazons. He speaks of their culture as being very real, of their characteristics as being fierce female hunter-warriors, repudiating men, and being a nomadic tribe.

AN AMAZONIAN ADVENTURE

Here's a story about the Amazons told by Herodotus. Many centuries ago, a group of Greek raiders journeyed into what we now call northern Turkey. Traveling across the steppe, the Greeks came across a group of warrior women. They kidnapped the women, imprisoned them in the holds of their ships, and set sail for home. But the Amazons escaped. They recovered their weapons and killed their captors. But, as they were horsewomen and not navigators, the ships drifted far off course. Ultimately, however, the Amazons landed in the Crimea, went ashore, and stole some horses. Once back in the familiar saddle, they began marauding, accumulating loot, and building up their strength.

Now, not far from where the Amazons came ashore, there dwelled a settlement of Scythians, generally a nomadic, horse-riding people of the steppe. But this settlement was different. These were so-called *royal* Scythians, affluent traders who had settled down in towns. The royal Scythians were wise. They didn't want to get raided by nearby commotion, so they sent out scouts, and the scouts found that the strange marauders were Amazons.

The Scythians were intrigued, so they ditched their scheme to send out soldiers to murder the marauders. Enter plan B: Assemble a party of nice young men. Sure, life in the settlement was luxurious, but the women of the royal Scythians tended to dwell indoors, away from the men. A bright lascivious idea was conjured by the royal Scythian men: Perhaps a few wild and fearless Amazons could spice up the town. So, the expectant band of bachelors journeyed out in search of the warrior horsewomen. They eventually came upon a single Amazon walking alone, and soon one became many, and the Scythians and the Amazons consolidated their respective camps.

The young men got excited, as young men are wont to do. They suggested the Amazons return home with them. To return to the settlement where they had houses and money and family. Was not a settled life superior to life on the steppe? The Amazons, amazed, begged to differ. Forget a settled life of drudgery. Leave town behind and live on the wild side with us; raiding, riding, and sleeping under the stars. Needless to say, the men packed their things, as young men are wont to do.

Ultimately, Herodotus reports, the Sarmatians, the people descended from the union of the Scythians and the Amazons, set up a society characterized by sexual equality. Women and men led the same sort of life. It's a story, a wonderful myth, in which the answer to the question of who will be dominated and tamed is "no one."

AMAZONS: MYTH OR REALITY?

This culture of the Amazons, from which Wonder Woman sprung—how true is it? What of the story of these warrior women? Myth or reality? In her book, *The Amazons: Lives and Legends of Warrior Women across the Ancient World*, Adrienne Mayor, a Stanford classicist and the world's leading expert on ancient female warriors, suggests that, while not totally true in all details, the tale is still broadly true.

The evidence, she says, is based on the fact that there really were warrior women. In some archaeological digs in Eurasia, up to 37 percent of warrior graves incorporate the bones and weapons of horsewomen who fought alongside men. In Mayor's words, "arrows, used for hunting and battle, are the most common weapons buried with women, but swords, daggers, spears, armor, shields, and sling stones are also found."

Incidentally, Mayor's own personal journey is a fascinating one. She studies the folklore, myth, and science of the ancient world. It is an obsession she has had for many years. She explains:

As a kid, I was a tomboy. I played with toy cowboys and Indians and soldiers and noticed there weren't any girls. Then, as a college student at the University of Minnesota, during the Vietnam War, I got interested in military history; I just thought that stories from wartime have the very best of human behavior and the very worst. I got permission to take R.O.T.C. (Army Reserve Officer's Training Corps) classes (back then, there weren't any girls allowed). Then I took classes in ancient Greek and Roman history. I was fascinated by the war stories about Amazons. In 1990, I proposed an article about Amazons to *Military History Quarterly*, and it was turned down. So I searched around, found a male coauthor, and had the guy propose the same article, and it was accepted.

These warrior women Mayor speaks of were those the ancient Greeks met on their expeditions around the Black Sea. And they triggered similar tales among travelers from ancient Egypt, China, Persia, among other locations. In the ancient Greek world, such women were objects of romantic intrigue. Their culture, in which both women and men were equal, acted as a counterpoint to Greek society, in which only men could be courageous. And so Greek tales about the Amazons, Mayor suggests, voice an ancient Greek appetite for gender equality.

The academic attitude to the Amazons has evolved. For a considerable time, according to Mayor, "most people argued that the Amazons on Greek vases were purely symbolic, that they represented, for example, young women who weren't yet married." But this view has recently been challenged by "a wealth of archaeological discoveries that show that there were women who behaved like Amazons, who wore the same clothes, who used weapons, who rode horses, and who lived at the same time as the ancient Greeks."

Due to such archaeological discoveries, we have a better picture of what Amazon life was like. The Amazons were probably Scythian nomads. They explored and wandered the territory north of the Black

Sea, between the Balkans, to the west, and the Caucasus, to the east. Contrary to vulgar myth, they weren't, as Mayor puts it, "man-hating virgins," but merely members of a culture "notorious for strong, free women." The Greek way of war focused on tactics associated with armored, brawny men. In contrast, Mayor writes "on the steppe the horse was the great equalizer, along with the bow and arrow, which meant that a woman could be just as fast, just as deadly, as a man."

THE LIFE OF AN AMAZONIAN

Life as an Amazonian was active and varied, to say the least. They spent days on end riding their horses. Amazonian life followed an annual cycle. In what reads somewhat like the "Strength through Joy" program of the Germans in the 1930s, the Amazons enjoyed large gatherings for feasting, athletic contests, and purifying saunas.

Ancient Greeks also credited the Amazons with inventing trousers. These, due to the biting cold out on the steppe, were worn by the Amazons along with long-sleeved tunics and pointed hats with earflaps. (With some imagination, it's the kind of look one can imagine on the catwalk.) The Amazons were tough in food and drink, too. They smoked marijuana, which is native to Central Asia, and drank fermented mare's milk, which they froze, then skimmed off the surface ice to increase the alcohol hit. And they were elaborately tattooed. The image of a Wonder Woman that one's mind conjures up from all these facts about the Amazons is less of the golden-eagled chest plate and star-spangled culottes or skirt, evolving later into the bottom half of a bikini, and more of a heavy inked Mongolian warrior of the female variety, looking more like an extra from *Mad Max: Fury Road* than the classic Wonder Woman. The Amazons appear to have domesticated dogs and hunted with eagles, so at least that much about the Wonder Woman origin story has some evidence behind it.

The Amazons were adaptable. When the need arose, they fought on foot. For example, one archeological research project of skeletal Amazonian head wounds from battle-axes suggested that the majority of blows were dealt by a right-handed opponent in direct face-to-face combat. Studies also suggest that there was fierce competition for resources and land. Little is known about their spiritual lives, or even if they had a spiritual side at all. But, as Adrienne Mayor writes, the archaeology and the folktales present us with "an impressionistic sense of the beliefs of the women archers of Scythian lands known as Amazons, an intangible mosaic of animism, totemism, magic, of sacred fire and gold, of reverence for Sun, Moon, sky, Earth, nature, wild animals, fantastic creatures. And horses." Little wonder that Marston was seduced into basing a superhero on such women!

HOW MYTHS AND SCIENCE FICTION WORK

Naturally, the facts of the lives of the Amazons were transformed into Greek myth. That's because, in the ancient Greek world, myths were a kind of thought experiment, a little like science fiction has been since the late 1800s. Look at what Mary Beard, professor of classics at Cambridge University said about Greek myths in an episode of the long-running BBC Radio 4 program, *In Our Time*: "You have to assume as a starting point that the fact that these things go on being told and recounted and sung and written about must mean they're doing a really important job." Professor Beard points out that this isn't some kind of "mad conservatism" on the part of the ancient Greeks. "I think myth is a very economical form of thinking about the world, as a set of ways of thinking about how the world is. The underlying function is to help us think about what human existence is like, and why it's so jolly difficult and hard, and why we do what we do."

Scholar Adrienne Mayor agrees that much of the ancient Greek mythmaking was centered around thought experiments. "What would happen if our Greek heroes encountered a band of Amazons? Sparks would fly!" Mayor also says that the original title of her book, *The Amazons: Lives and Legends of Warrior Women across the Ancient World*, "was going to be *Amazons in Love and War*, because there were just as many love stories as there were war stories." But the love stories differ from each other in crucial ways. "In the stories that the Persians and the Egyptians told, they were often attracted to the women they were fighting: their impulse was, we want them on our side, we want them as companions and lovers." For example, in one of her favorite myths, told in Iran, Egypt, and elsewhere, "a prince fights a warrior princess; they're so equally matched that the fight goes on and on, and when they sit down to rest, they fall in love."

By contrast, the ancient Greeks had a "uniquely dark mythic script: all Amazons must die, no matter how attractive, no matter how heroic." Mayor suggests that the Greeks admired the Amazons more than any other enemy. The Amazons are never portrayed in Greek art as taking flight from peril or begging for mercy, but lasting love between Greek men and Amazons was always depicted as impossible. As Mayor puts it, "every Amazon that we hear about in Greek mythology is heroic; heroes who are the equals of the greatest male Greek heroes." She holds that in ancient Greek portrayals of Amazons, "you can discern some yearning and desire for some kind of resolution to the tension between 'Yes, we want them as our companions,' and 'We couldn't possibly, because we have to control our own women.'" And yet, in another twist, the Amazons had a special place in the lives of Greek women. "Amazons were featured everywhere, on women's pottery, on perfume jars, on jewelry boxes, on sewing equipment. Little girls played with Amazon dolls." It's a peek, according to Mayor, into "a mystery of Greek private life." It seems like an ancient Greek version of modern kids playing with superhero figures.

A LITTLE SLINKY?

We can easily see why William Marston chose the ancient Greek world and the Amazons for his backstory on the origins of Wonder Woman, but the Amazonian image of an assertive superwoman proved too much for some sections of conservative American society. Wonder Woman was first presented to the world toward the end of 1941 in *All-Star Comics*, and, at the beginning of 1942, on the front cover of a brand-new comic book, *Sensation Comics*. Marston's chosen cartoonist for drawing Wonder Woman was artist Harry G. Peter. Peter had much experience. He had drawn not just the hourglass-figured, pouty-mouthed Gibson Girls (the personification of the feminine ideal of physical attractiveness), but also suffragist artwork of women breaking out of their oppressive chains. From the get-go, Wonder Woman was founded in feminist advocacy, the idea that women and men were equal in worth, and the opposition to the subordination of women. As a sign of the times, Wonder Woman left paradise to fight fascism with feminism in the land of the brave and the free. "America, the last citadel of democracy, and of equal rights for women!"

The early Wonder Woman comics were a space where gender stereotypes were confronted. This was hugely helped by the subversive fluidity created during WWII, where the wartime threw up feminist exemplars such as Rosie the Riveter and others, who were portrayed as working for the war effort. Similarly, Diana, Wonder Woman, is super in strength and confidence. And Diana's plucky sidekick and best friend, Etta Candy, with whom she shares many adventures, is not only fearless and capable, but also a key factor in Diana's successes.

Together with their friends, Diana and Etta battle Nazis, Japanese spies (portrayed dubiously with darker skin and very large teeth), and myriad mythical and alien characters. Their MO is not overt violence but conciliation, working especially to redeem the female perpetrators, and generally come to the aid of women and children. When Diana is taken prisoner, she often releases herself, or else Etta and her friends come to the rescue. These women are single-minded. Showing little inclination for mere romance, their focus is fun and fighting injustice, a considerable step up from the contemporary expectations for women.

The sisterhood of Diana, Etta, et al. mimics the camaraderie of the Amazons, and sits in stark relief to more recent portrayals of women as combative and catty. Importantly, Diana takes the world of women along with her. She is not the typical exceptional superhero. Wonder Woman is strong and capable, sure. But so are *all* women. Witness the fact that, should Diana submit to mere men, it is her undoing. For, if a man chains Diana's bracelets together, "she becomes weak as other women in a man-ruled world." Diana often vocalizes lessons in independence and strength. She tells other women they can be the same as long as they believe in themselves and have the necessary physical training. Her own formidable strength is often on display, lassoing planes, throwing cars, smothering bombs in her hands, and campaigning for just causes, such as standing up for female workers in a department store to get "a living wage."

While it might have seemed like good, clean, progressive, and patriotic fun to some, it was too much for others. And by March of 1942, the *National Organization for Decent Literature* placed *Sensation Comics* on a blacklist of "Publications Disapproved for Youth." The reason? Wonder Woman was not sufficiently dressed, as cartoonist Harry G. Peter chose to represent Wonder Woman as sporting a tiara of gold, a bustier in red, panties of blue, and knee-length, red leather boots. But there were greater grounds for complaint. In February of 1943, Josette Frank, a scholar of children's literature, sent Max Gaines a letter. She confessed that she'd never really been a Wonder Woman fan, but now she felt she had to speak up about Wonder Woman's "sadistic bits, showing women chained, tortured, etc." Josette Frank had a point. Tale after tale, Wonder Woman is either lassoed, fettered, and manacled or gagged, bound, and chained. The Amazon even cries at one point, "Great girdle of Aphrodite! Am I tired of being tied up!"

WAY TOO KINKY

These depictions were no accident. For example, in a tale about Mars, the God of War, in his original scripts, William Marston gave artist Harry G. Peter incredibly specific instructions for the serial art in which Wonder Woman is taken prisoner:

> Close-up, full-length figure of WW. Do some careful chaining here; Mars's men are experts! Put a metal collar on WW with a chain running off from the panel, as though she were chained in the line of prisoners. Have her hands clasped together at her breast with double bands on her wrists, her Amazon bracelets and another set. Between these runs a short chain, about the length of a handcuff chain—this is what compels her to clasp her hands together. Then put another, heavier, larger chain between her wrist bands which hangs in a long loop to just above her knees. At her ankles show a pair of arms and hands, coming from out of the panel, clasping about her ankles. This whole panel will lose its point and spoil the story unless these chains are drawn exactly as described here.

Moreover, and later in the same story, Wonder Woman is imprisoned. Struggling to listen to a conversation in the next room, through the amplification of "bone conduction," she clasps the chain between her teeth. Marston's instructions ran, "Close-up of WW's head shoulders. She holds her neck chain between her teeth. The chain runs taut between her teeth and the wall, where it is locked to a steel ring bolt."

CHAIN OF FOOLS?

When Gaines realized they had a problem with all this Wonder-Woman-in chains stuff, he forwarded Josette Frank's complaints to Marston, but Marston paid little heed until Dorothy Roubicek, the first female editor at *DC Comics*, also made it known that she objected to Wonder Woman's torture. Marston mansplained to Gaines. "Of course I wouldn't expect Miss Roubicek to understand all this. After all, I have devoted my entire life to working out psychological principles. Miss R. has been in comics only six months or so, hasn't she? And never in psychology." Marston then made it far worse by adding that "the secret of woman's allure" is "women enjoy submission, being bound."

But Roubicek had far more experience than Marston credited her for. She had also worked on Superman and invented kryptonite. She thought that superheroes, male or female, should have their limits and vulnerabilities. Roubicek educated Gaines: Wonder Woman should be equal to Superman. Just as Superman wasn't able to return to the planet Krypton, Wonder Woman shouldn't be able to return to Paradise Island, where most of the kinky stuff seemed to happen.

Gaines sent Roubicek to interview Lauretta Bender, an associate professor of psychiatry at New York University's medical school, an expert on aggression, and a senior psychiatrist at Bellevue Hospital, where she was director of the children's ward. Roubicek reported back. Professor Bender did not "believe that Wonder Woman tends to masochism or sadism." Bender also approved of the way Marston was "playing" with feminism.

> She believes that Dr. Marston is handling very cleverly this whole "experiment," as she calls it. She feels that perhaps he is bringing to the public the real issue at stake in the world (and one which she feels may possibly be a direct cause of the present conflict) and that is that the difference between the sexes is not a sex

problem, nor a struggle for superiority, but rather a problem of the relation of one sex to the other.

Roubicek's conclusion was simple: "Dr. Bender believes that this strip should be left alone."

THE LIST OF MENACES

For the time being, Marston was off the hook, and Gaines was partially reassured. But the issue arose again in September 1943 when Gaines got a letter from US Army staff sergeant John D. Jacobs, saying "I am one of those odd, perhaps unfortunate men who derive an extreme erotic pleasure from the mere thought of a beautiful girl, chained or bound, or masked, or wearing extreme high heels or high-laced boots, in fact, any sort of constriction or strain whatsoever." Sergeant Jacobs went on to inquire whether the author of Wonder Woman himself had possession of any of the sex aids described in his tales. Gaines sent Sergeant Jacobs's letter on to Marston, with an alarmed covering note, "This is one of the things I've been afraid of."

It was time for action, so Gaines sent, for Marston's information, a memo penned by Roubicek that comprised a "list of methods which can be used to keep women confined or enclosed without the use of chains. Each one of these can be varied in many ways; enabling us . . . to cut down the use of chains by at least 50–75 percent without at all interfering with the excitement of the story or the sales of the books." Marston was unabashed. He replied by return of post, "I have the good Sergeant's letter in which he expresses his enthusiasm over chains for women; so what?" Marston claimed that, as a practicing clinical psychologist, he was unconcerned.

Some day I'll make you a list of all the items about women that different people have been known to get passionate over;

women's hair, boots, belts, silk worn by women, gloves, stock-
ings, garters, panties, bare backs. You can't have a real woman
character in any form of fiction without touching off a great
many readers' erotic fancies. Which is swell, I say."

Marston was confident that he knew what the real red line was.
Innocuous erotic fantasies are one thing, he claimed, but it's "the
lousy ones you have to look out for, the harmful, destructive, morbid
erotic fixations; real sadism, killing, bloodletting, torturing where the
pleasure is in the victim's actual pain, etc. Those are 100 percent bad
and I won't have any part of them." Finally, he added, "please thank
Miss Roubicek for the list of menaces."

During those war years, comic books boomed. On average, two
to three comics were avidly consumed by over 90 percent of six- to
eleven-year-olds each week. Among twelve- to seventeen-year-olds the
figure was 84 percent, and for those readers eighteen and older it was
35 percent. Perhaps surprisingly for those who bow to the cliche of
comic books being the sole reserve of the male geek, the male:female
ratio of those reading comics was roughly even.

Precise circulation numbers are pretty tricky, but *Wonder Woman*'s
success was such that its first title, in *Sensation Comics*, was launched
by the publisher into franchises in *All Star Comics* and *Comic Cavalcade*
too. When Wonder Woman then became syndicated to newspapers, the
general sales manager for the King Features Syndicate produced this
pitch for the newspapers: "It is a strip that will appeal to women for the
strength and character of Wonder Woman; to men for her magnificent
beauty and her sinuous agility, her daring exploits and her charm; to
all boys for the sheer thrill and excitement of Wonder Woman; and to
girls who will take her for their idol and prototype." In all likelihood,
Wonder Woman at this time was probably selling in the region of six,
and maybe even seven, figures.

WONDER WOMAN GETS SYNDICATED

It was toward the end of WWII, when the fight against fascism had finally turned a positive corner, when Gaines and Marston signed *Wonder Woman* over to become a newspaper strip. As Marston was now busy with this new strip, he drafted in support to help him write comic-book scripts. That support came in the shape of eighteen-year-old student, Joye Hummel. Hummel's tales were more innocuous than those of Marston. In her own words, the editor at DC would "always okay mine faster because I didn't make mine as sexy." In celebration of this new syndication, an artistic panel was commissioned in which Superman and Batman are shown rising out of the front page of a daily newspaper, declaring "Welcome, Wonder Woman!" as she is pictured leaping onto the page.

The political context of the *Wonder Woman* chains controversy was the backstory of the fight for female rights. Marston and Harry Peter had both been hugely influenced by the feminist suffrage movement. And the suffrage movement had used chains as the crux of its iconography. When Marston had been a student at Harvard, the college had received British suffragist Emmeline Pankhurst as a guest speaker on campus. Pankhurst had achieved huge fame as a freedom fighter for chaining herself to the gates outside 10 Downing Street, home to the British Prime Minister and headquarters of the government since 1735.

And when American feminist Margaret Sanger faced charges of obscenity for her scientific account of birth control in the magazine she founded called *Woman Rebel*, a petition was sent to President Woodrow Wilson on her behalf. It ran, "While men stand proudly and face the sun, boasting that they have quenched the wickedness of slavery, what chains of slavery are, have been or ever could be so intimate a horror as the shackles on every limb, on every thought, on the very soul of an unwilling pregnant woman?" American suffragists threatened to chain themselves to the gates outside the White House.

Chains figured in other protests too. For example, in Chicago in 1916, suffragists from the states where women still hadn't gotten the right to vote marched in chains. And in Sanger's 1920 book, *Woman and the New Race*, Sanger argued that woman "had chained herself to her place in society and the family through the maternal functions of her nature, and only chains thus strong could have bound her to her lot as a brood animal."

THE TIDE TURNS

Many superheroes didn't survive long after the war ended. And those that did abide in peacetime were changed for good in 1954. That's when German American psychiatrist Fredric Wertham published his book, *Seduction of the Innocent*, arguing that comic books negatively influenced children. Wertham then testified before a Senate subcommittee set up to investigate comics. Wertham went further. He held that comics were corrupting American kids. They were being slowly turned into juvenile miscreants.

Wonder Woman was a target of particular dislike for Wertham. Bender had earlier written that *Wonder Woman* tales display "a strikingly advanced concept of femininity and masculinity" and that "women in these stories are placed on an equal footing with men and indulge in the same type of activities." Wertham, however, found the feminism in *Wonder Woman* repellant: "For boys, Wonder Woman is a frightening image. For girls she is a morbid ideal. Where Batman is anti-feminine, the attractive Wonder Woman and her counterparts are definitely anti-masculine." According to Wertham, Wonder Woman "is physically very powerful, tortures men, has her own female following, is the cruel, 'phallic' woman. While she is a frightening figure for boys, she is an undesirable ideal for girls, being the exact opposite of what they are supposed to want to be." Wertham went on to ask "what are the activities in comic books which women 'indulge in on an equal footing with men?' They do not work. They are not homemakers. They do not bring up a family. Mother-love is entirely absent. Even when Wonder Woman adopts a girl there are Lesbian overtones."

Professor Bender begged to differ. She too testified at the Senate hearings, defending Wonder Woman. If any part of modern American culture was bad for girls, Bender contested, it certainly wasn't Wonder Woman. It was Walt Disney. "The mothers are always killed or sent to the insane asylums in Walt Disney movies." But Bender lost the debate.

THE CRIMES OF McCARTHY

In Wertham's secret papers, housed at the Library of Congress, and only made available in 2010, there is a striking professional hostility to Bender. The background to this antagonism is that Bender's late husband had been Wertham's boss for some considerable time, so the antipathy appears more professional in nature than anything to do with the content of the comics. For instance, Wertham's papers include a piece of scrap paper upon which he listed "paid experts of the comic book industry posing as independent scholars." First on the list of the comic book industry's lackeys was Bender, about whom Wertham noted, "boasted privately of bringing up her three children on money from crime comic books."

The political context of these Senate hearings was McCarthyism. In the early 1950s, the American political class constantly informed the public that they should fear various subversive communist influences in their lives. The reds could be lurking anywhere. They could cunningly be using their positions as teachers, professors, union leaders, poets, comic book writers, artists, or journalists to serve the mission of "world communist domination." Hell, even the janitor couldn't truly be trusted. This paranoia, which is exactly what it was with regard to the perceived communist threat from within, and which is often referred to as the "Red

Scare," hit its fevered apogee between 1950 and 1954. That's when right-wing Republican Senator Joe McCarthy of Wisconsin launched a spate of hugely publicized probes into alleged communist infiltration. Nowhere was safe. Up for interrogation was not only the State Department, the White House, and the Treasury, but also the US Army. During Dwight Eisenhower's first couple of years in office, McCarthy's hysterical denunciations and scaremongering whipped up a climate of fear and suspicion across the US. In scenes reminiscent of the witch hunts of the past, few dared tangle with McCarthy for fear of being labeled, as McCarthy wrote in 1953, "any man [sic] who has been named by a either a senator or a committee or a congressman as dangerous to the welfare of this nation, his name should be submitted to the various intelligence units, and they should conduct a complete check upon him. It's not too much to ask."

It's in the midst of these "witch hunts" that we should view the 1954 Senate hearings involving *Wonder Woman*. Nonetheless, after the hearings, DC Comics sacked Bender from her role on its editorial advisory board. And the Comics Magazine Association of America wrote itself a new code: comic books could no longer contain anything cruel: "All scenes of horror, excessive bloodshed, gory or gruesome crimes, depravity, lust, sadism, masochism shall not be permitted." Nor could the comics include anything kinky: "Illicit sex relations are neither to be hinted at nor portrayed. Violent love scenes as well as sexual abnormalities are unacceptable." And, finally, anything deviant or unconventional was to be thrown right out the window. "The treatment of love-romance stories shall emphasize the value of the home and the sanctity of marriage." It was a victory for conservative America.

STEPFORD DIANA

The end of WWII brought other changes. The more flexible gender roles of the wartime Wonder Woman came to a close through the offices of one Robert Kanigher, who became *Wonder Woman*'s editor and writer for the next two decades. Meanwhile, William Marston, physically weak by 1945, died in 1947, and with Marston died the will to focus on feminism in *Wonder Woman* comis. Post-war conservative America was busy with reactive policies on gender roles, forcing women out of work and back into the home, to become the supportive spouses and mothers they were allegedly born to be.

It's in the shadow of this reaction that comics turned away from superwomen and toward a twofold obsession, first with sickly romance, where women were shown as only desiring of heterosexual relationships to complete them. (Ira Levin's famous 1972 novel, *The Stepford Wives*, which has been adapted into several films, is a superbly satirical "feminist horror" of such sickly romances. Levin's tale follows Joanna Eberhart, a successful photographer and mother of two who moves with her husband and children from New York City to the idyllic town of Stepford, Connecticut. But Joanna soon finds that the women in Stepford are all weirdly submissive and conformist. They have no interests or aspirations of their own. As Joanna tries to uncover the mystery behind such female transformation, she realizes that the men of Stepford have replaced their wives with robots designed to be submissive and obedient. In the end, Joanna discovers that she too has been targeted for transformation and must fight for her life to escape the fate of the other Stepford wives.) And indeed, straight-up horror was the second obsession of post-war comics in conservative America, tales in which women were portrayed as attractive victims.

WONDER WOMEN ARE HISTORY

A yearslong moral panic over comics culminated in the industry-adopted Comics Code, which took aim at violent content in comic books, as well as the "nontraditional" nature of prominent genres and characters (remember that Wertham expressed concern that Diana's all-female entourage was gay and that Batman and Robin were gay, casting them all outside of American Cold War–era convention.)

The days of Robert Kanigher began with serial artists Irwin Hasen and Ross Andru giving Diana a makeover fit for the Princess of Themyscira. Her utilitarian boots were ditched for more delicate laced sandals. Her previously flat bustier was recast with a decidedly more defined curvature. Diana's hair was longer, her eyes and mouth bigger, and her now more spirited posing helped readers focus on her feminine curves.

Kanigher's *Wonder Woman* of the 1950s and 1960s centered on "traditional" femininity, masculinity, and marriage. Marriage proposals to Diana became constant, if not tedious, and Wonder Woman abandoned the military to work as a stunt woman in Hollywood while also writing a romance column. Artist Harry G. Peter was sacked, and *Sensation Comics* canceled. "Wonder Women of History," a feature at the back of *Wonder Woman* comics that profiled famous women, was replaced with "Marriage a la Mode," a feature highlighting marriage customs across the globe.

In short, post-WWII Wonder Woman evolved from being progressive and antiestablishment to rather conservative, if not reactionary. For instance, her initial backstory, in which Diana was "made from clay by her mother and brought to life by goddesses," becomes a volatile memory which is dumped in favor of her mother confessing tearfully that, in fact, Diana's father was lost at sea (heaven forbid the notion that men be made redundant in procreation), and Wonder Woman's mission to "teach peace and equality" was ousted by the more conservative undertaking to "battle crime and injustice."

DIANA'S DEVOLUTION

Diana's powers also evolved. They became considerably more "super." Her abilities are now, it seems, gifted by the gods, instead of being earned through Amazonian training. The change is profound. There's a stark contrast between the elitist idea of being chosen from above and the far more leveling idea of a self-made marvel, an example that other women could easily ape. Indeed, the new, conservative Diana has scrapped Etta and the girls in favor of silly stories of Wonder Woman's past, a past in which she was a "Wonder Tot," then a "Wonder Girl," welcoming the romantic attentions of the likes of superior suitors like Mer-Boy and Amoeba-Boy while attending birthday parties, flying kites, and no doubt reinventing the wheel. Befitting of Diana's new-found elevated status, there is even an "upgrade" in her assailants, as she battles with increasingly ridiculous monsters: a giant squid, a giant bird, a giant centipede (being giant is clearly key here), and even the risible mustachioed Egg Fu, a Chinese communist agent who is perplexingly shaped like a *giant* egg the size of a house with a casually racist Charlie Chan–type speech pattern who uses his mustaches as whips against his enemies and was presumably based on the supervillain, criminal genius, and mad scientist of 1930s cinema, Dr. Fu Manchu.

In the same way that Diana is portrayed as a Wonder Tot and Wonder Girl at younger ages, as an adult she is now a doting mother caring for her "Wonder Family." And thus her devolution is complete. With the feminizing costume change and body morphing, with the constant romancing of the eligible Diana, and the sudden appearance of a "necessary" father, her resulting Wonder Family is the final brush stroke to the picture they wish to paint. The changes conjure up a woman back where she belongs; within a heteronormative framework. Consequentially, Diana no longer represented a refutation of the kind of social inequalities as she had in the 1940s. As with many other American institutions, comic books in the Cold War championed an ideology which worked to reinstall bourgeois order after the many challenges of the war years and the fight against fascism. (Ironically, of course, policies and portrayals tending to the control of media, an obsession with national security, crime and punishment, and sexism, with society portrayed as male dominated with traditional gender roles and the state as sponsor and the supreme guardian of the family institution, are *all* characteristics of fascism!)

Such changes were not popular among the comic book readership. Fan letters, published in the comics themselves, grew ever more disgruntled, so Kanigher tried returning to a more "golden age" Diana veneer. The revision wasn't to everyone's liking. One reader expressed concern on the letter page, though admittedly the letter pages were typically manipulated by the editors, voicing concern about Kanigher's new direction and wondering whether the Wonder Family survives such changes, and suggesting that the character was now held in "extremely low opinion" by the fans. Kanigher penned a defiant response: "I want to see whether it is possible to recreate a golden era for a golden super heroine . . . If my head rolls for it, I'll try to stick it back on and move in the direction you've mentioned."

But Kanigher's recasting came without the golden era womanism of Marston. Subsequently, for the duration of the 1960s, sales were second-rate. And yet, throughout this evolution, Diana's intro mantra endured: "Beautiful as Aphrodite, wise as Athena, swifter than Mercury, and stronger than Hercules." The changes to come would dial up the Aphrodite, and dial down the triumvirate of Hercules, Mercury, and Athena.

KEEPING UP WITH BATS AND LEATHER-CLADS

Two of the most influential television programs of the 1960s were ABC network's *Batman*, which ran for three seasons from January 12,

1966, to March 14, 1968, and ITV's *The Avengers,* a British espionage television series created in 1961, which ran for 161 episodes until 1969. The former featured Batgirl, and the latter Emma Peel. Both programs were influential on the portrayal of Diana. On the covers of the 1968 issues of *WW* #178 and #179, Diana's hair is now longer and teased up, and she is pictured in a purple tunic with Emma Peel–style black leggings and boots: "Wow, I'm gorgeous! I should have done this ages ago!" Diana also uses martial arts, gadgets, and detective skills to best the criminal fraternity. Indeed, Superman observes as Diana takes out some muggers, "Guess I was wrong worrying about Diana! The lass doesn't need superpowers!" Moreover, during the day, Diana owns and runs a clothing boutique. She is her own boss, and financially independent.

The main writer and later editor of this era was Dennis O'Neil, who went on record to define the new direction for *Wonder Woman* as more feminist: "I saw it as taking a woman and making her independent, and not dependent on superpowers. I saw it as making her thoroughly human and then an achiever on top of that [which was] very much in keeping with the feminist agenda." Once more, the influence of *Batman* and *The Avengers* is clear. The year previous, the new Batgirl on the *Batman* TV show was portrayed as an attractive working woman *sans* superpowers but nonetheless fighting crime, and, of course, owned by the same company in DC.

WONDER WOMAN ON TV

Perhaps the most memorable portrayal of Diana from the sixties and seventies was the *Wonder Woman* TV series, which aired for three seasons between 1975 and 1979. The first season starred Lynda Carter as the titular character Wonder Woman/Diana Prince and was set, like the original comic book, in the 1940s. The second and third seasons, as the first season was so expensive to produce as a period piece, were set in then-current day late 1970s, with a title change to *The New Adventures of Wonder Woman*.

As a result of the TV series, the character of Wonder Woman underwent a huge explosion in popularity from the relative doldrums of the conservative fifties and sixties. Lynda Carter's Diana Prince was rounded and versatile. She was strong. She was smart and confident. She was fierce and sexy, yet humble and diplomatic. She could also be funny and sweet. Hell, she was even regal (but one shouldn't perhaps hold that against her). All in all, the show was seen as pushing the envelope of conventional narratives of gender on prime-time TV.

The first season of the program harked back to Marston's comic strips of the 1940s set during WWII. It depicted not only Diana but also her mother, sister, and other Amazons, as well as Etta. On occasion, Diana scolded criminals with sermons on peace, equality, and tolerance, but the show also had plenty of live action, including Carter's iconic "spin" that magically transformed her identity and appearance without the need to dive into a telephone box (going one better than Clark Kent).

A couple of season openers also portray Prince as the victor in a contest of strength and skill, allowing her to leave the all-female Paradise Island and return Steve Trevor to the US, though the series didn't go as far as to reference Diana's birth from clay. Elsewhere, Diana reaches out to other women with calls to sisterhood, most famously in the first season's second episode, "Fausta, the Nazi Wonder Woman." Wonder Woman falls through a false floor where a Nazi gang overpowers her with chloroform. The unconscious Wonder Woman is kidnapped and carted off to Nazi Germany for study. Self-important Steve launches his mission to rescue Wonder Woman but exacerbates the situation by being captured himself shortly after. Meanwhile, Wonder Woman orchestrates her own escape and has to return to Nazi Germany to rescue Steve and convince Fausta to abandon the Nazi cause with appeals to sisterhood. Indeed, Steve is knocked unconscious or requires rescue in many other episodes but is "reprogrammed" to be more appreciative of Wonder Woman's alter ego Diana Prince.

Sadly, once more, both the TV show and its associated comic declined in sales and audience ratings. The show became more like a 1970s cop show, with Diana more like a supersleuth and the only continuing female character. One female fan wrote in to the comic to complain: "You said Wonder Woman wasn't selling very well . . . In one year it has gone from my favorite comic to worst and there are many reasons . . . She has no friends and no life . . . there are no supporting characters. Even the villains are forgettable."

By 1981, sales of the *Wonder Woman* comic were half of what they were in the late seventies while the TV show aired. And in that period, of the late seventies and early eighties, the plot portrayals evolved into a campy and action-tedious affair, with women either fighting one another or else set against men in a kind of cliched "battle of the sexes" manner. In the last issue of *Wonder Woman* before a company-wide reboot of its comics, Diana and Steve get wed, suggesting the event to be Wonder Woman's sole dream come true.

One letter writer in 1986 summed up the past two decades of *Wonder Woman* comics:

She should be one of the most successful, interesting, thought-provoking characters in comics. Instead, we have seen her presented as everything from a female Superman . . . to a man-hating belligerent female chauvinist. We have seen her stripped of power and we have seen her prove that Diana Prince can be a hundred times more boring than Clark Kent. We have seen her comic format changed to WWII to fit a TV show.

WONDER WOMAN ON THE SILVER SCREEN

Clearly, the depiction of Wonder Woman has changed over the years. Her evolving image is no doubt due to a complicated nexus of the contemporary political milieu, comic-book audiences, and changing writers and illustrators. The more feminist and less sexualized portrayals of the 1940s gave way to the devolved heteronormative, warlike, and more sexualized depictions for much of the 1950s and 1960s. A differential ebb and flow between these two poles has persisted ever since. What might the future bring?

Wonder Woman's premiere on the big screen was in *Batman v Superman: Dawn of Justice*, where she was portrayed by Gal Gadot. The fact that this Israeli actress and model was once crowned Miss Israel 2004, then represented her country at the Miss Universe 2004 pageant, *then* served in the Israel Defense Forces for two years as a combat fitness instructor seems to suggest that the screen image of Diana is tending to heteronormative and warlike. Indeed, the movie's costume designer, Michael Wilkinson, commented that the film's director Zach Snyder wanted Wonder Woman "to be a fierce and intimidating warrior—gritty, battle-scarred, and immortal."

In 2017, *Wonder Woman* became the first female-led superhero movie since *Catwoman* in 2004. The movie was greeted with enthusiasm by *The Guardian*:

As a woman, it's . . . a warm swell of relief that she is such a glorious badass, one who wears her femininity with the same pride and poise that she wears her armor-plated bra . . . The masterstroke of this origins story . . . is that it accentuates and celebrates Diana's feminine traits. Her secret weapon is not the bullet-repelling jewelry; not her swordplay; not her ability to fire shockwaves from her wrists. It's not even her luxuriant, swooshing hair. It is her empathy. Although now I mention it, the hair *is* pretty impressive.

THE ELUSIVE FEMALE SUPERHERO

As we have seen over the last seventy-five years, the ideation of Wonder Woman, that of a female warrior who embodies power and compassion, capability and heroism, equality and empathy, has so often not been matched with her depictions in mainstream media. There are those among her many fans who feel that DC isn't really sure how to deal with Diana. The challenge of Wonder Woman being so "tricky" to write, along with the evolving changes in her portrayals, could explain why there have been so many evolutions of this superwoman.

Yet, it's not Diana herself that's so damned tricky to write. Consider Diana's origin story, along with her powers and core values. They can be summed up in one sentence: "Super strong Amazonian warrior, molded of clay, birthed by goddesses, travels to the world of men to teach peace, love, understanding, and equality." Daunting, right? But, on further inspection and thought, no more daunting than "airborne, solar-powered, and faster-than-a-speeding-bullet, super strong man from a dying alien world journeys to Earth as a baby, grows up in Kansas, and fights for justice." So, not tricky at all. The truth of the matter is that we live in a rather tricky and unequal world—one in which women are still devalued. If all us humans were viewed as equal, a superwoman wouldn't be seen as tricky, as she wouldn't be seen as threatening the social order through her very existence.

CHAPTER 5

WHY IS SHURI SUCH
A BIG DEAL?

(. . . in which we explore the various social, cultural, and educational factors in the public's perceptions of cinema scientists)

The Draw-A-Scientist Test (DAST) is an open-ended projective test designed to investigate children's perceptions of the scientist. Originally developed by David Wade Chambers in 1983, the main purpose was to learn at what age the well-known stereotypic image of the scientist first appeared. Following the simple prompt, "Draw a scientist," 4,807 primary school children in three countries completed drawings. The drawings were then analyzed for seven standard indicators: lab coat, eyeglasses, facial hair, symbols of research, symbols of knowledge, products of science (technology), and relevant captions. From these indicators, Chambers was able to show that children began to develop stereotypical views of scientists from a very early age, with a progressively larger number of indicators appearing as the grades progress.

—David Wade Chambers, *Science Education* (1983)

FEMALE SCIENTISTS IN THE MOVIES

What kind of image is conjured up in your imagination when you think of a "scientist?" For most people, still, it's most likely a brainbox wearing a white lab coat and sporting round-rimmed glasses, a little like Sheldon in *The Big Bang Theory* or Professor Frink in *The Simpsons* or Dr. Henry Wu in *Jurassic World*. Yes, very probably, you imagine the scientist as a male. Scientists in the movies rather unimaginatively conform to the character type most people have in their heads. Thankfully, in recent decades, moviemakers have demanded something a lot more imaginative, calling for scripts which reverse spin the archetype.

Consider some of the best female scientists in film. In following list, by no means exhaustive, numerous and different areas of science are represented: there's a paleobotanist, an astrobiologist, a primatologist, a xenobotanist, an astrophysicist, a biologist, a medical engineer, and a linguist!

1. Let's begin with paleobotanist Dr. Ellie Sattler (played by Laura Dern), who first appeared in 1993's *Jurassic Park* film. Rather than scream and scarper in the tradition of old Hollywood, and due to Dr. Ellie Sattler's athleticism, she survives numerous dinosaur attacks and helps in the survival of others.

2. Astrobiologist Dr. Ellie Arroway's (Jodie Foster) search for extraterrestrial life is front and center in the movie version of Carl Sagan's spectacular extraterrestrial drama, *Contact*. Dr. Arroway is not just brilliant but also passionate about science, making the first contact with an alien civilization and persuading her government to build a machine based on plans beamed out by the extraterrestrials. She's also courageous enough to risk her life by traveling in the machine to witness the wonders of deep space.

3. Sigourney Weaver became so famous for her feminist role as Ellen Ripley in *Alien* that she was handpicked to play female scientists in two other influential movies. In the 1998 film, *Gorillas in the Mist*, Sigourney Weaver plays real-life primatologist Dian Fossey, a champion for a rare species of mountain gorillas in Rwanda, who put her life at risk to protect the gorillas from poachers and potential extinction. *Gorillas in the Mist* was nominated for five Academy Awards, and Weaver received a nod for Best Actress.

4. Sigourney Weaver also played xenobotanist and head of the Avatar Program, Dr. Grace Augustine, in James Cameron's 2009 blockbuster, *Avatar*. Dr. Augustine is a no-bullshit boffin who penned the book on plant life on Pandora, though the doctor clearly enjoys the company of plants over people. (By the way, the word "boffin" is an informal label, mainly British, which is given to scientists who know much about science, but little about ordinary life; those things that make us more human.)

5. Natalie Portman plays the famous female role of astrophysicist Dr. Jane Foster in the Thor movies. Without Foster's technical help, Thor would have remained powerless during his exile on Earth. Sure, there's a little of the old-fashioned romance about their relationship, but that spark between Foster and Thor is the catalyst that triggers the superhero into eventually learning a little more humility and humanity from the exemplary scientist.

6. Anne Hathaway plays NASA biologist Dr. Amelia Brand in Christopher Nolan's epic cosmic drama, *Interstellar*. Brand is assigned the nearly impossible task of being the chief scientist on a team that must travel through a wormhole in order to save the human race.

7. Sandra Bullock plays NASA medical engineer Dr. Ryan Stone in the 2013 sci-fi drama, *Gravity*. The doctor courageously commits to a half-year training program, as she is responsible for the invention of new medical imaging technology. Though not an astronaut, Dr. Stone attempts to fix a space shuttle while in orbit, struggles to survive against all odds in the extreme environment of outer space, and though battered, bruised, and exhausted, she somehow finds the will to live.

8. Finally, Amy Adams plays linguist Dr. Louise Banks in the 2016 film, *Arrival*. The movie's plot revolves around Dr. Banks, who is the key scientist in the very first human communicative contact with extraterrestrials. The movie makes a point of focusing on a feminist theme through the telling of the tale of Dr. Banks and the way in which she balances the demands of not only being a professional scientist but also a loving mother.

But what do all the aforementioned female scientists have in common? They're all white. And that makes our final female scientist, Shuri, in the critically acclaimed *Black Panther* movie, stand out all the more. Like Wonder Woman, Shuri, sadly, is a princess. But don't let this "Disneyfication" put you off. Shuri is also a young, black, female scientist who excels in STEM as a brilliant and ingenious tech whiz and inventor. She is the brains behind harnessing the power of vibranium, particularly in the creation of the famous Black Panther suit. Let's take a look back at the way scientists used to be portrayed in cinema, take stock of the progress that's been made, and try to understand why Shuri is such a big deal.

MALE, MAD, AND DANGEROUS

For a long time, all the scientists in the movies were male and either dangerously mad or absurdly saintly. What's worse is that the mad male boffins (the fictional ones made-up as part of the movie's plot) have outnumbered the saintly ones (the real-life scientists portrayed in biopics) by a huge margin. For example, a glance through some six thousand movie reviews from *Variety* up to 1993 shows that fewer than thirty have been depictions of real-life scientists. Many, many more have been biopics of real-life artists, or celebrities, or even, God help us, politicians. During the twentieth century, this bias fed into the school system. *Famous People on Film*, the 1977 exhaustive guide to the short films or documentaries used in schools, colleges, and communities in the UK, includes just seventeen scientists out of almost one thousand total entries. Most of these boffins are predictable: the physicists Albert Einstein, Isaac Newton, and Galileo Galilei; chemists Alfred Nobel and Louis Pasteur; biologist Charles Darwin; inventors Alexander Graham Bell and Thomas Edison; and that most celebrated of shrinks, Sigmund Freud. The list also includes a trio of pioneers associated with the development of domestic electricity, the Italian Alessandro Volta, the Frenchman André-Marie Ampère, and the German Georg Ohm (British engineer James Watt was not on the list as, perhaps confusingly, he had little to do with light bulbs.)

It's been a similar story in the states. After 9/11, a three-minute film was compiled featuring a selection of clips from 110 movies celebrating "the essence of the United States." It was part of the healing, a way of reassuring people about the future. Interestingly, the compilation included not a single biopic of American scientists, inventors, or discoverers (though it *did* include the 1957 Charles Lindbergh film, *Spirit of St. Louis* and the 1983 space-race drama, *The Right Stuff*.) And yet the compilation saw fit to include Kubrick's mad *Dr. Strangelove*, *The Nutty Professor*, and two "dirty science" movies in 1979's *The*

China Syndrome (featuring a faulty nuclear reactor and corrupt corporations) and 2000's *Erin Brockovich* (industrial pollution). The previously positive "can do" slant of 1930s and 1940s biopics of US boffins and inventors, what Disney himself termed "stick-to-it-ivity," was no longer in vogue.

SCIENCE AND HORROR

Moreover, consider the following in-depth study of more than one thousand horror films distributed in the UK between the early 1930s and 1980s showing that mad male scientists or their creations had been the monsters/villains of 31 percent of the total, or 41 percent if we include mutations. Furthermore, in these horror movies, scientific or psychiatric research was responsible for 39 percent of the threats in all plots, compared to just 11 percent of "natural" threats. What about the idea of scientist as hero, you might wonder? Sadly, scientists have been the heroes of only 10 percent of horror films, and most of those were before the 1960s. When the stats are broken down, the figures show that male psychotics took over from mad male scientists in the period between 1960 and 1984, a "golden age" of horror when a remarkable 74 percent of all horror movies were made.

These are worrying trends. First, the fact that science in one form or another was the main organizing hypothesis of horror movies from that era. Second, that pre-1960 horror films are dominated by science, and post-1960 horror by sex and psychos. And third, that the decline in the number of scientist-heroes correlates to a decline in the probability of the expert being met with success, and a concomitant rise in the role of the "every-person" novice as hero. In horror movies, qualifications counted for nothing. This trend reflects the rise of the horror movie. In the early 1930s, horror films accounted for 1 percent of total film distribution. This figure remained pretty constant until the mid-1950s, when it rose to 1.3 percent. By 1957, the figure was 6.9 percent; by 1973, it rose to 9.1 percent; and by 1983, it was 12 percent. Ever since, horror movies have become mainstream, one of Hollywood's principal products, in the way Westerns used to be.

LE CINÉMA ET LA SCIENCE

Horror notwithstanding, how has the rest of cinema portrayed science and the scientist? A fascinating survey on this was conducted by Professor Jacques Jouhaneau for the centenary of cinema publication, *Le Cinéma et la Science*, published in 1994. The survey used a wide-ranging approach, which confirmed the decline of the mad male scientist and the rise of the, mostly male, "scientists with interesting private lives." *Le Cinéma et la Science* reported that, out of sixteen thousand movies of all genres (fantasy, comedy, drama, adventure, science fiction), distributed between the dates of 1910 and 1990, and originating in numerous cultures, a full 520 movies were studied in detail, and these films featured 560 key characters who were professors, doctors, researchers, or scientists.

Data for the 520 movies was analyzed against three axes: the reputation of the scientist in relation to society (the range here runs from -10 for "murdering psychopath" to +10 for all-conquering real-life hero); the authenticity of the science portrayed and the likelihood of its existence (the range here runs from -10 for the likes of *The Invisible Man* to +10 for real-life experimenters); and the date of the movie's release. The most common science disciplines represented by scientists in these films are biology, nuclear physics, psychiatry, informatics, and archeology and exploration.

On the axis of the scientist in relation to society, most movies (221) lie within the range +7 to +5. This grouping of science "heroes" is designated as an Indiana Jones type: active adventurers, defenders of society, or witnesses. The second biggest grouping, a total of nine-ty-seven, lies within the range +4 to +2 and is Doc Brown-designated: victims of their passions and weaknesses. The third most populous group, with a total of sixty-eight, sits in the -5 to -7 range, and is given a kind of Dr. Frankenstein designation, being described as mad experimenters, megalomaniacs, and makers of monsters. The most criminal Hannibal Lector kind of category, confirmed murderers and serial killers, accounts for a total of nineteen movies.

Over time, there has been a notable shift from the "mythic savant with dangerous and derisory ideas" to the "researcher whose social function is recognized, whose science is within the bounds of pos-sibility—even while encouraging contradictory responses." In short, an evolution from the mad male scientist to the well-intentioned but powerless male scientist.

INSANE AND MALE: CINEMA'S MAD DOCTORS

The heyday of the mad male doctors was the 1930s. And boy did they make their mark. It began with Dr. Rotwang, the evil genius at the center of Fritz Lang's hugely important and pioneering sci-fi block-buster of 1927, *Metropolis*. We get a little more detail about the story of arguably cinema's all-time most influential scientist in Thea von Harbou's original novel of 1925, which was the basis for, and written in tandem with, Fritz Lang's futuristic film. Von Harbou painted a tantalizing picture:

> There was a house in the great Metropolis which was older than the town. Many said that it was older, even, than the cathedral . . . Set into the black wood of the door stood, copper-red, myste-rious, the seal of Solomon, the pentagram . . . It was said that a magician, who came from the East (and in the track of whom

the plague wandered) had built the house in seven nights . . . Parchments and folios lay about, open, under a covering of dust, like silver-grey velvet . . . Then came a time which pulled down antiquities . . . but the house was stronger than the words on it, was stronger than the centuries. It hardly reached knee-high to the house-giants which stood near it. To the cleanly town, which knew neither smoke nor soot, it was a blot and an annoyance. But it remained.

Fritz Lang's sci-fi movie masterpiece was produced in Germany at the height of the Weimar Republic. It was the most expensive silent movie of its day, and so its influence was huge. The style adopted for Lang's film has been described as "ray-gun gothic." The film's design elements were based on contemporary modernism and art deco, with its dark, futuristic dystopia of skyscrapers and brooding tensions between social classes.

Set in the middle of this ray-gun gothic, like the very eye of a brewing storm, was Rotwang, the most significant of mad male doctors. The persona of Rotwang and his old house was an early-in-the-movie announcement of Lang's central belief that "an audience learns more about a character from detail and décor, in the way the light falls in a room, than from pages of dialogue." It's the same kind of recipe used for the portrayal of mad doctors to come.

Rotwang, closely aided and abetted by the likes of Dr. Frankenstein, begat a bevy of insane progeny in the cinema of the 1930s, such as Bela Lugosi's Dr. Mirakle in the 1932 movie, *Murders in the Rue Morgue*, a film based very loosely on the Edgar Allan Poe tale. Dr. Mirakle is a carnival sideshow entertainer and scientist who kidnaps Parisian women to mix their blood with that of his gorilla, Erik. (I'm not kidding!) Also in 1932 was the movie *Dr. X* (Elon Musk would approve). In the plot of *Dr. X*, there is a bizarre cannibalistic serial killer on the loose who kills only when the moon is full. Police track this killer to a scientific institute, where every one of the employees is a mad male scientist. Each one is a suspect, and each one is disfigured in a different way. (More about mad doctor disfigurement soon.)

Yet again in 1932 (what a year it was for mad male doctors), there was *The Island of Lost Souls*. Cinema's best ever version of H. G. Wells's *The Island of Doctor Moreau*, *The Island of Lost Souls* hosts legendary British actor Charles Laughton as Dr. Moreau, a mad scientist who creates human-like hybrid beings from animals via vivisection, and simply doesn't care how many creatures he tortures, maims, or kills to achieve his goals.

A mad and dark amalgam of Rotwang, Frankenstein, and Moreau conjured up the likes of Dr. Fu Manchu, another archetype of the evil criminal genius scientist, who featured in cinema, television, radio, comic strips, and comic books for over ninety years; Lex Luthor, nemesis of Superman who's been portrayed as both a power-hungry CEO and a narcissistic and egotistical mad scientist; and Dr. Strangelove, Stanley Kubrick's dark creation of 1964, the sinister Nazi scientist, a heady fusion of RAND Corporation strategist Herman Kahn, Nazi-SS-officer-turned-NASA-rocket-scientist Wernher von Braun, and Edward Teller, known as the father of the hydrogen bomb.

The portrayal of the mad male scientist continued into the 1960s. Consider Doctor Julius No in the 1962 Bond movie, *Dr. No*. He is a German-Chinese malevolent boffin, whose two prosthetic metal hands, due to radiation exposure, pile on the evil scientist tropes. The two black gloves he wears are key to the mad character. It has long been the case that moviemakers use motifs to signify something slightly strange or other about a mad scientist. Cary Grant is made to wear Coke-bottle-bottom glasses to play a mild-mannered paleontologist way back in 1938's movie, *Bringing Up Baby*. His specs are an exterior sign of interior issues—they imply that the scientist doesn't see things like the rest of us. His myopic condition also cuts him off from some element of the real world. In this way, the filmic choice of a simple prop symbolizes separateness from the emotional depth that the rest of us, hopefully, enjoy, rather than the obsessive focus of someone whose vision is limited to that of pure intellect alone.

The backstory of Dr. No's black-gloved hands first crystallized with Lang's *Metropolis*, for Rotwang had also possessed the devilish detail of a metal hand (a relic of some dark accident that left him maimed), an emblem of his struggles with dark science, but also of his twisted mental powers. This kind of recipe was taken on board in the portrayal of later cinema scientists with unnamed scars, withered limbs, crazy eyes, curiously outsized foreheads (think Doc Brown), and, naturally, the wheelchair. This cliche becomes quite ridiculous at times. Take the scientist figure in *Mad Max Beyond Thunderdome*, for example. This character is a dwarf, sitting on the shoulders of a giant (probably a reference to Issac Newton's famous disingenuous quote, "if I have seen further, it is by standing on the shoulders of giants"). In *Mad Max*, the brain has become detached from the body, as it were, yet Newton had been referring to the shoulders of fellow scientists such as Descartes, Galileo, and the ancients rather than some ripped jock on whose shoulders he could be carted about. This cinematic trail of cliches leads directly to the door of Doctor Julius No. Dr. Strangelove had just the one gloved hand, which spastically jerked into a Nazi salute at the mere mention of the word "slaughter," but Dr. No managed *two* black prosthetic hands. How badass is that? The hands, an outcome of some dubious nuclear accident, gift Dr. No some uncanny powers.

Uncanny is a curious word in science. Sigmund Freud used the title, "The 'Uncanny,'" for his 1919 essay about the specific psycho-logical fears associated with missing parts of the human body, writing "Dismembered limbs, a severed head, a hand cut off at the wrist . . . feet which dance by themselves—all these have something peculiarly uncanny about them, especially when, as in the last instance, they prove capable of independent activity in addition." In the 1935 movie, *Mad Love*, Peter Lorre plays a mad doctor obsessed with the female star of the Grand Guignol theater in Paris. The lady thanks him for his attentions but tells him she is retiring from the stage to live with her concert pianist husband. But the husband is caught up in a train accident in which he loses his hands. Lorre's mad doctor saves the husband, but now "gifts by transplant" the hands of a murderous knife thrower.

While it's true that Freud's examples are drawn from a more ancient fantasy associated with witchy folk tales, his analysis still applies to modern fantasies, such as science fiction. The black-gloved hands of Doctors Strangelove and No conjure up infantile fears, and rather unhomely histrionics. After all, the German word "unhomely" is also used for "uncanny." For Freud, "the severed hand has a particularly uncanny effect." And so we see the same game at play with Dr. Rotwang, whose metal hand replaced his severed one, with Dr. Strangelove, whose whole arm is now "capable of independent activity," and finally with Dr. No, whose hands are replaced by metal prostheses in the movie.

Even though the actual portrayal of the scientist has greatly evolved, the same conventional cliches are found throughout popular culture: the scientist is shown as a misfit, single-mindedly obsessed with his (very rarely, her) work, detached from society in general and often from the scientific establishment in particular.

MAD MALE COMIC DOCTORS

Just look at the way in which the image of the scientist was then developed in comic books: a lunatic parade of lone-genius mad male doctors, working mostly in isolation, wreaking havoc on the modern world. The depiction of these nutty professors belongs to another long-gone day:

the original mad doctor in Frankenstein, based partly on an age-old Elizabethan tale in Christopher Marlowe's play, *Doctor Faustus*, and the terrible genius of Rotwang. Even when science fiction stories moved on from these dark origins of a singular maverick scientist, private capital was almost always the new king in town: Wayne Enterprises or Stark Industries, LexCorp or Oscorp, Pym Tech or Roxxon.

Yet, when you take a closer look at these company portfolios, they're still dominated by mad male doctor bosses. Wayne Enterprises, a company that sells pretty much everything, is run by a tech-obsessed absentee CEO who battles with Gotham crime dressed as a flying mammal. Stark Industries, a corporate brand that took a market mauling when its battery-powered CEO involuntarily created an army of murderous bots, raised an entire European city into the sky. The CEO of LexCorp, a company with huge stocks of radioactive kryptonite, is maniacally and fetishistically fixated on the destruction of a single alien. Pym Technologies is a brand far too reliant on the market viability of ant-talking tech, and Roxxon is a company in which many of the former CEOs languish in jail.

It's undoubtedly the case, of course, that positive and realistic portrayals of scientists may be hard to reconcile with the demands of effective dramatic representation. Especially comedy. Alternatively, with the increasing multiplication and atomization of popular culture, with hundreds of satellite channels, video games, and the internet, you would have thought we could do a little better and loosen the stranglehold that the stereotypical images of science and scientists have on popular culture. It seems not. But things *have* improved. A little.

EVER SINCE ROTWANG

After the rather inauspicious start for scientists in the cinema, negative portrayals of science persisted between 1940 and 1945, somewhat understandably, and the late 1970s and early 1980s. Of these negative movie science characters, 41 percent were doctors, 35 percent unspecified savants (be they researchers, inventors, creators, or problem solvers), and 26 percent psychologists and psychoanalysts. The outright "dangerous" characters with no scruples slowly evolved into sorcerer's apprentices for corporations or professors with sexual hang-ups.

The conclusion to *Le Cinéma et la Science* is clear:

The cinematic scientist is, above all, a "hero" of the male sex. His scientific activities are, most often in recent times, in the domain of the plausible. Generally devoted to activities which have huge public interest, he can however content himself with being a simple, straightforward witness of his time, called in to resolve everyday problems . . . His sharpness of mind, however, does not exclude weaknesses of the flesh. His evolution over time inexorably expels him from the abstract domain of "absolute knowledge"—however loosely defined—toward the domain of "methodical doubt" authorized by increasingly real scientific knowledge. In parallel, the savant who is mad, cursed, or homicidal continues on his way—but he becomes much more stereotyped than his heroic adversary, to whom he serves merely as a standard of value. As for his activities, they belong to the imaginary realm *and are unlikely to generate in the mind of the spectator a really negative image of science* [my emphasis].

SCIENCE HAS AN IMAGE PROBLEM

The generational parade of mad male doctors at the movies has had its effect on the public image of science. Yet it's not *all* the fault of cinema. Here's what Theodore Roszak had to say about the stereotypes of science in his 1974 work, *The Monster and the Titan: Science, Knowledge, and Gnosis*:

> Asked to nominate a worthy successor to Victor Frankenstein's macabre brainchild, what should we choose from our contemporary inventory of terrors? The bomb? The cyborg? The genetically synthesized android? The behavioral brainworker? The despot computer? Modern science provides us with a surfeit of monsters, does it not? I realize there are many scientists—perhaps the majority of them—who believe that these and a thousand other perversions of their genius have been laid unjustly at their doorstep. These monsters, they would insist, are the bastards of technology: sins of applied, not pure science. Perhaps it comforts their conscience somewhat to invoke this much-muddled division of labor . . . Dr. Faustus, Dr. Frankenstein, Dr. Moreau, Dr. Jekyll, Dr. Cyclops, Dr. Caligari, Dr. Strangelove. The scientist who does not face up to the warning in this persistent folklore of mad doctors is himself the worst enemy of science. In these images of our popular culture resides a legitimate public fear of the scientists' stripped-down, depersonalized conception of knowledge—a fear that our scientists, well-intentioned and decent men and women all, will go on being titans who create monsters. What is a monster? The child . . . of power without spiritual intelligence.

In other words, not only is the mad male scientist so ingrained in the modern psyche, but we also often don't even realize he's there. And the real debate to be had should not be about creationism versus Darwinism, or whether artificial intelligence is good or evil, but about the role of science in late-capitalist society. After all, the darker side of invention is also the responsibility of science: the atom bomb, gunpowder, plastics, climate change, refined sugar, Agent Orange, and many more. And as Roszak puts it, "The scientist who does not face up to the warning in this persistent folklore of mad doctors is himself the worst enemy of science."

AN INTERLUDE ON THE EQUALITY OF WOMEN IN EVOLUTION

Consider this interview between British journalist Michael Parkinson and Polish-British mathematician and philosopher, Jacob Bronowski about Bronowski's epic thirteen-part 1973 BBC television documentary and accompanying book, *The Ascent of Man*:

Parkinson: "Let's talk about something then that I know featured in *The Ascent of Man*, and that's this thing about equality between men and women. It's fascinating reading your book which has been written about the series. I didn't realize it, in fact, as a species, physically, we are closer together than any other; are we not? Men and women. I mean, for instance, I didn't know that a woman in our species is the only female in any species to have an orgasm, which is extraordinary. I didn't know that. And we are the only species that copulate face to face."

Bronowski: "I ought, as usual, to say *almost* the only species, in both cases. There are some aquatic mammals, like the whales and the seals, that find the other form [of intercourse] inconvenient, and it is also true that it has recently been shown that, in some of the bigger monkeys, the females would be able to achieve orgasm but don't in the wild because the males don't keep at it long enough. You'll pardon me for cutting the scientific jargon and stating the plain facts."

Parkinson: "But where does that leave us, then, in the situation with the so-called equality, or inequality, of the sexes. I mean, what conclusion do you arrive at?"

Bronowski: "I think it's the most powerful thing that can possibly be said. *It is quite clear that in the human species there is less difference between male and female than in any related species.* For instance, you can tell a female chimpanzee from a male chimpanzee instantly, because one weighs, you know, about twice what the other does. You can tell a female gorilla from a male gorilla instantly, and their whole relationship is dominated by these physical differences. Human beings are extraordinarily alike, and that's not just a matter of the fashions of the last ten years. They have always been very much alike by comparison with their related species, that is the large monkeys. Moreover, they are much more alike in their emotions. You spoke about women having orgasms. But the fact is that women are almost the only creatures among the mammals that are sexually receptive at all times, as males are, that therefore do not come on heat as most other animals do, and that, in general, enjoy a position of intellectual equality, emotional equality, with men, and that's been going on for over two million years; that didn't start with women's lib [women's liberation movement]. Now, I take that a little further. You're quoting from the end of program twelve, which I happen personally to be in love with, because I think that those few lucid phrases about women say so much more than much else of the science in the programs. Women exercise as much choice in the male that they're going to mate with as do men. That's rare among animal species. And it means that, from an early age, the evolution of the human race has been mediated by the fact that women have looked for special skills in men, and that men have had to be able to provide those skills, for instance, skill of hand, manipulation of the hand by growth of the brain, which have made the human race come on at such a tremendous rate."

"WHEN I THINK ABOUT A SCIENTIST, I THINK OF . . . ?"

Bronowski and actual scientific evidence notwithstanding, Hollywood images can be very powerful. And to show just how powerful they can be, American cultural anthropologist Margaret Mead conducted numerous studies of Western schoolchildren to find out how they tended to draw or describe images of scientists, and the cultural shorthand they used. Mead's pioneering studies began in 1957 with a survey of thirty-five thousand American high school children.

The context of Mead's mission was the Cold War. In particular, the fear felt by America's national scientific associations and foundations about low levels of recruitment and the peril of falling behind the Russians. Mead asked the kids to complete the statement, "when I think about a scientist, I think of . . . ?" and take it from there. The contemporary and collective image of the scientist, when Mead began her study, came out in the following fashion:

The scientist is a man who wears a white coat and works in a laboratory. He is elderly or middle-aged and wears glasses. He is small and stout or tall and thin . . . He may wear a beard, may be unshaven and unkempt. He may be stooped and tired. He is surrounded by equipment: test tubes, Bunsen burners, flasks and bottles, a jungle gym of blown glass tubes and weird machines with dials . . . He spends his days doing experiments. He pours chemicals from one test tube into another. He peers raptly through microscopes. He scans the heavens through a telescope. He experiments with plants and animals, cutting them apart, injecting serum into animals. He writes neatly in black notebooks.

We can see that the cultural tropes, traits, and expectations of the "scientist," always male, are commonly shared. Upon this foundation is then based a number of divergencies. And, within those divergencies, any positive traits tend to be expressed at a distance, as it were, from the child's own life or career trajectory:

He is a very intelligent man—a genius or almost a genius . . . He is careful, patient, devoted, courageous, open-minded. He knows his subject . . . He works for long hours in the laboratory, sometimes day and night, going without food and sleep. He is prepared to work for years without getting results and face the possibility of failure without discouragement; he will try again. One day he may straighten up and shout: "I've found it! I've found it!" He is a dedicated man who works not for money or fame or self-glory, but for the benefit of mankind and the welfare of his country. Through his work people will be healthier and live longer, they will have new and better products to make life easier and pleasanter at home, and our country will be protected from enemies abroad. He will soon make possible travel to outer space.

MEAD DISHES THE DISAPPOINTMENT: ALL ABOUT "HE"

Far more ominously, the negative side of the scientist's image was very much at the forefront of the survey form:

The scientist is a brain. He spends his days indoors, sitting in a laboratory, pouring things from one test tube into another . . .

Though he works for years, he may see no results or may fail, and he is likely to receive neither adequate recompense nor recognition. He may live in a cold-water flat; his laboratory may be dingy. If he works by himself, he is alone and has heavy expenses. If he works for a big company, he has to do as he is told; he is just a cog in a machine. If he works for the government, he has to keep dangerous secrets; he is endangered by what he does and by constant surveillance and continual investigations. If he loses touch with people, he may lose the public's confidence—as did [J. Robert] Oppenheimer. If he works for money or self-glory he may take credit for the work of others . . . He may even sell secrets to the enemy. His work may be dangerous. Chemicals may explode. He may be hurt by radiation, or may die . . . He may not believe in God or may lose his religion.

In short, the scientist is a male brain that is so involved in his work that he doesn't know what's going on in the world.

He has no other interests and neglects his body for his mind . . . He neglects his family—pays no attention to his wife, never plays with his children. He bores his wife, his children, and their friends with incessant talk that no one can understand; or else he pays no attention or has secrets he cannot share. He is never home. He is always reading a book. He brings home work and also bugs and creepy things.

Margaret Mead came to a shocking conclusion. Even the positive cinematic image of the scientist ultimately proved unattractive for young Americans. They rather respected the scientist than wanted to actually become one. The main factor in this state of affairs? The "mass media image of the scientist," for which cinema bore a special responsibility. Mead lamented, if only mass media could focus on the following:

> Real, human rewards of science—the way in which scientists today work in groups, share common problems, and are neither "cogs in a machine" nor "lonely" and "isolated." And if only tales about scientists could dissolve "the sense of discontinuity between the scientist and other men . . . to bring about an understanding of science as part of life, not divorced from it.

Mead felt that a good start would be to stop talking about "*the scientist, science,* and *the scientific method,* and to use instead the names of the particular sciences." In Mead's day, late 1950s America, negative stereotypes precisely matched Hollywood's movie output and the associated media of network television.

THE "DRAW-A-SCIENTIST" TEST

Mead's survey was all about "he," not "she," and hardly anyone heeded it. Two-and-a-half decades later, research scholar David Wade Chambers completed an eleven-year survey on whether Mead's results still held true. Chambers also wanted to work out, if little had changed since Mead's day, at what approximate age the stereotypes first made their mark in the minds of children. Chambers's survey centered on a "Draw-a-Scientist" Test to 4,807 schoolkids in the 5–11 age range; mostly situated in Montreal, Quebec. Each drawing was analyzed according to seven basic indicators of "the standard image of the scientist," which he extrapolated from Mead. But gender still wasn't one of them.

1. Lab coat (most often white)
2. Spectacles

3. Facial hair (!)
4. Symbols of research: scientific instruments and laboratory equipment
5. Symbols of knowledge: mostly books and filing cabinets
6. Technology: the by-products of science
7. Relevant captions: formulae, taxonomies, the "eureka!" syndrome, etc.

The conclusion of the Chambers study was this: Kindergarteners were too young to have accumulated any of the aforementioned indicators. By the second and third grades, roughly the ages of seven and eight, the male scientist stereotype starts to take shape, with roughly two to three indicators per child, and commensurate with Aristotle's famous contention, "Give me a child until he is seven and I will show you the man." By the fifth grade, ten years of age, most of a class is likely to show at least four indicators, while some exhibit as many as six or seven. Most shockingly of all, only girls drew women scientists. And out of a total of over *two thousand* girls, only twenty-eight drew female boffins.

Mad male doctors made their way into Chambers's study too. Besides the seven basic indicators, in statistically significant numbers appeared "clear representations of the Jekyll/Hyde and Frankenstein legends, magical portrayals of alchemical laboratories, the frightening visions of clearly deranged (often labeled 'mad') scientists testing, for example, new improved versions of the electric chair." These representations also included Bond-like lairs of underground laboratories; the presence of guns, bombs, and even armed missiles; and a plethora of signs on lab doors and walls declaring "Keep Out," "Top Secret," "Do Not Enter," and "Go Away!"

The general conclusion of the Chambers study was this: The stereotypical image of the scientist that emerged among high school students in the Margaret Mead survey was still alive and kicking among grade school students a generation later. Moreover, various "mad doctor" motifs were added to it as students progressed through the grades.

ENTER THE CYNICAL CORPORATE MALE SCIENTIST

Chambers also referenced further surveys showing that this cliched stereotype of the scientist being male had changed little since the start of the twentieth century. The only evolution had been the relative demise of the individual "mad doctor" and the advance of the cynical corporate scientist. Out with the rolling eyes and drooling mouth and in with "a new sanitized professional image" with an increasing emphasis on government science, secrecy, and warfare as a result of lab-coated scientists seen in printed and televised product advertisements.

The conflicts at the core of this stereotype endured. On a more positive note, the (mostly male) lab-coated scientists, either bald or sporting a Doc Brown frizz and wearing Coke-bottle eyeglasses, are working alone indoors or underground in their laboratory lairs marked "secret;" they are obviously detached from everyday life; they are middle-aged and physically unattractive. On the more negative note (not that the positive is *all* positive!), the survey respondents understand that progress in science and tech has made everyday life a lot better, and that doing science can be an engaging, important, and exciting ("eureka!") job, even if it's not one they would choose for themselves.

NEW SCIENTIST SURVEY

Similar surveys of adult readers have shown that such stereotypical images are no way restricted to schools: they are utterly embedded in the general culture. For example, a concerted *New Scientist/New Society* questionnaire, reported in the *New Scientist* in August 1975, led to contradictory conclusions. Sure, the survey's scientist respondents knew that their colleagues were normally "approachable, sociable, open, unconventional; as having many interests and as being popular." But non-scientist respondents viewed them as the reverse, "remote, withdrawn, secretive, and conventional. Scientists have few interests and are rather unpopular."

Large majorities in both respondent groups continued to view "the scientist" as male. In response to the prompt, "when I think about a scientist, I think of . . .?" the majority of lay respondents replied with the name of a real-life scientist: In order of popularity, Archimedes, Darwin, Galileo, Newton, Louis Pasteur, Linus Pauling (though not Crick and Watson), and bouncing bomb engineer Barnes Wallis. Almost all of these real-life scientists had been played in movies. For the, perhaps unlikely, record, ancient Greek mathematician, physicist, and engineer, Archimedes, had actually "starred" in Giovanni Pastrone's 1914 Italian silent film epic, *Cabiria*, which had Archimedes, as is legend, focusing the sun's rays via a giant reflecting glass at the Roman fleet during the battle of Syracuse, thereby torching its sails.

A similar and maybe more likely starring role for Archimedes happened more recently in 2023's *Indiana Jones and the Dial of Destiny*, in which a tortured plot is spun involving Archimedes, an Antikythera mechanism, and time travel! Bomb engineer Barnes Wallis had, of course, starred in the 1955 WWII classic, *The Dam Busters*, whose depiction of the RAF's 617 Squadron 1943 attack on the Möhne, Eder, and Sorpe dams in Nazi Germany was the inspiration for the Death Star trench run in 1977's *Star Wars: Episode IV—A New Hope*.)

Non-lay respondents, meanwhile, often tellingly wrote, "when I think of scientists, I think of my colleagues," instead of naming some celebrity scientist. The assumption underlying the survey question, and of the lay replies, was that scientists work in isolation, and that the science professions are best embodied by great male discoverers.

THE WELLCOME TRUST MORI POLL

In a similar vein to the studies of Mead and Chambers was the large-scale MORI poll of more than 1,500 working scientists' attitudes, carried out by the Wellcome Trust from late 1999 to March 2000. The poll concluded that even though 97 percent of the scientists surveyed saw huge benefits in the public having a greater understanding of science, with 70 percent accepting that their research had social and ethical implications, a full 61 percent felt that they couldn't really do much about it themselves due to their day-to-day on-the-job pressures and the pervading culture of publish or perish.

Around roughly the same time (February 2000), the UK's House of Lords Select Committee published a report concluding that "a new culture of dialogue between scientists and the public" was needed. While that was predictable enough, parts of the Wellcome survey were far more telling:

Many scientists still have a low opinion of the public's ability to understand science, are distrustful of most of the media, and believe they themselves are viewed in an equally negative light by the public . . . While scientists are most likely to view themselves as enquiring, intelligent, and methodical, they typically believe that the public sees them as detached, poor at public relations, secretive, and uncommunicative . . . Just over one in three scientists (35%) considered the media to be a barrier. Only one in five considered the lack of communication skills among scientists themselves to be a barrier to improved understanding of science among the public. Although the nonspecialist public is believed to rely most heavily on the mainstream media to keep themselves informed about scientific issues and their implications . . . scientists in fact place little trust in any information source other than those provided by their peers. Nevertheless, the media was [sic] seen to be something of a necessary evil; 73% of respondents considered it to be the most effective method of communication with the public.

HATS OFF TO SHURI

Considering the findings of Margaret Mead's 1957 survey, the *New Scientist*'s 1975 survey, and D. W. Chambers's 1983 survey, it's clear that stereotypes of scientists remained a major part of the culture of the post-war period, and scientists were reluctant to engage with that culture. And the whole thing was complicated by the scientific community's poor regard for the lay ability to understand science, and for mass media's role in its presentation of science.

In stark contrast to the surveys discussed above is the presentation of Shuri in Black Panther media. Portrayed in the MCU movies by Letitia Wright, Shuri is, of course, the sister of T'Challa, the Black Panther, and Princess of Wakanda, but far more importantly she is the leader of the Wakandan Design Group, which is key to the creation of many of Wakanda's technological innovations. In short, Shuri is a scientist and inventor at heart.

Shuri is also the mastermind behind all the extremely advanced and vibranium-related tech that comes out of Wakanda, which is way ahead of most other contemporary technology. The energy-absorbent suit, the remotely piloted car, the 3-D holograph communicator—all her design. But Shuri is not just intelligent, she is also a courageous and social creature. Rather than remaining in her lab, Shuri plays an executive and crucial role in the execution of the tech missions she invents. She drives the getaway car. She battles hand-to-hand with full-fledged rebels. She even subverts all the cliches about what a hero's younger sister should be. In short, Shuri is a young, intelligent, determined, brave, and heroic character in her own right.

For way too long, the cinema scientist has been mostly male and mostly "mad." Shuri is powerful, intelligent, and fearless; a beacon of import to millions of young girls. In her article for *Vanity Fair*, Johanna Robinson wrote, "After a packed advance screening of *Black Panther* in Los Angeles last week, two young boys went bounding ahead of the crowd leaping for joy and punching the warm night air. They weren't pretending to be Black Panther, or even another Wakandan warrior. They were pretending to be Shuri."

CHAPTER 6

HOW DID SUPER-BARBIE GO FROM PLASTIC TO PROFOUND?

*(. . . in which we peel back the feminist layers
of Greta Gerwig's film)*

Barbie has answered the billion-dollar question with a resounding "yes." Barely three weeks into its run, writer-director Greta Gerwig's blockbuster has raked in an astounding $1.03 billion at the global box office, according to official Warner Bros. estimates. This makes Gerwig the first solo female director with a billion-dollar movie. As one half of the viral "Barbenheimer" phenomenon, it isn't shocking *Barbie* has performed well. And, standing on her own two feet, the doll's incredible success is not unexpected at all. "I've been in this game for thirty years and the Barbie and 'Barbenheimer' phenomenon is as unprecedented as it was unpredictable," said Paul Dergarabedian, senior media analyst at Comscore . . . According to Dergarabedian, only about fifty films in history, unadjusted for inflation, have hit the billion-dollar mark. He added the movie's marketing campaign was the first hint *Barbie* would be a box-office smash. "The marketing campaign for Barbie set into motion a chain of events that led to the word 'Barbenheimer' being added to the popular lexicon by virtue of its shared release date with *Oppenheimer*, and that's when we all knew something very special and unique was going to create a much bigger than expected outcome for the film not only for the opening weekend, but for its global run in theaters."

—Eva Rothenberg, CNN, "Barbie makes history with $1 billion at the box office" (2023)

AN UNCONVENTIONAL SUPERWOMAN

The movie blockbuster hit of summer 2023 was about a peculiar kind of superwoman. Originally a fashion doll created by American business-woman Ruth Handler, and made by American toy company Mattel in 1959, more than a billion Barbie dolls have since been sold. While not exactly a conventional superhero, Barbie has become a cultural icon with an array of bestowed honors that is very rare in the world of toys.

Back in 1974, a section of New York City's Times Square was renamed Barbie Boulevard for a week. In 1986, legendary pop artist Andy Warhol created a painting of Barbie, which sold at auction at Christie's in London for $1.1 million and led to the Andy Warhol Foundation teaming up with Mattel in 2015 to create an Andy Warhol Barbie. In 2016, *Musée des Arts Décoratifs* at the Louvre in Paris held a Barbie exhibit, which featured seven hundred Barbie dolls over two floors, as well as multimedia works by contemporary artists that contextualized the cultural impact of the Barbie phenomenon.

In an article in late 2002, British weekly newspaper, *The Economist*, stressed the importance of Barbie to children's imagination:

From her early days as a teenage fashion model, Barbie has appeared as an astronaut, surgeon, Olympic athlete, downhill skier, aerobics instructor, TV news reporter, vet, rock star, doctor, army officer, air force pilot, summit diplomat, rap musician, presidential candidate (party undefined), baseball player, scuba diver, lifeguard, firefighter, engineer, dentist, and many more . . . When Barbie first burst into the toy shops, just as the 1960s were breaking, the doll market consisted mostly of babies, designed for girls to cradle, rock, and feed. By creating a doll with adult features, Mattel enabled girls to become anything they want.

BARBIE TURNS "SUPERHERO"

Barbie became an official superhero in 2015. Kind of. It took *that* long for Mattel to begrudgingly admit that girls also like superheroes. As Mattel put it in *Toy News Online*, "Barbie in *Princess Power* arrives in answer to a recent survey conducted by Mattel that revealed that nine out of ten girls around the globe 'wished there were more superheroes for girls.'"

So, Mattel recast their flagship doll into a superwoman. Now, when some poor soul meets a little difficulty, Barbie transforms from a pink-adorned blonde into a "super" pink-adorned blonde with special skills. For example, here are some of the superhero acts: fixing a skateboard; fixing her friend's hair; actually stopping time so one of her friends can carry out her chores in time to "later" hang out with a third friend. Notwithstanding the suspension of space-time itself, it's not a particularly impressive set of super-skills. But Mattel has an excuse for this, too, again quoted in *Toy News Online*, claiming that "the hybrid nature of the new range is a result of the finding that more than half of those girls asked disagreed that 'superheroes are more fun than princesses.'" In response, Michael Shore, PhD, vice president and head of global consumer insights at Mattel, said "Interestingly, girls' responses were split as to whether superheroes were more fun than princesses and we know girls like to explore multiple roles when engaged in imaginative play . . . Unlike other female superheroes, Barbie is the only one designed through the eyes of girls."

FIRST FEMALE MOVIEMAKER TO GROSS $1 BILLION

Making Barbie into a superhero was a good idea, but Mattel's effort, seemingly dreamt up by a boardroom full of males, came nowhere near justifying the idea. Enter Greta Gerwig. American actress, screenwriter, and director, Gerwig first grabbed collective attention while working within the mumblecore movie subgenre. This subgenre of independent movies typically centers on naturalistic acting and conversation and low-budget production with a focus on dialogue rather than plot and an emphasis on the relationships between young adults.

Post mumblecore, Gerwig collaborated with her husband, Noah Baumbach, on movies such as 2010's *Greenberg* and 2012's *Frances Ha*, for which she received a Golden Globe Award nomination. Half a decade later, as a solo moviemaker, Gerwig hit the heights again writing and directing the coming-of-age movies *Lady Bird* (2017) and *Little Women* (2019), both nominated for the Academy Award for Best Picture. For *Lady Bird*, Gerwig also received nominations for Best Director and Best Original Screenplay, and for *Little Women*, Best Adapted Screenplay.

BARBIELAND

So, what did Greta Gerwig *do* with Barbie? When the news first dropped that Gerwig was working on the *Barbie* project, a huge anticipation grew way before folks really knew what to expect. It wasn't as if Barbie came with much cultural baggage or prepackaged narrative, unlike some other superwomen from the Marvel or DC Universes. And that left Gerwig with the opportunity to cinematically recast the candy-colored feminist fable in a way that simultaneously celebrates, parodies, and deconstructs its happy-plastic subject in Barbieland.

Gerwig's masterpiece was described in *The Guardian* as a movie "reinvention of Mattel's most (in)famous toy . . . like a sugar-rush mashup of Pixar's *Toy Story 2*, Carlo Collodi's *Pinocchio*, the cult live-action feature *Josie and the Pussycats*, and the Roger Ebert–scripted exploitation romp *Beyond the Valley of the Dolls*." The story starts with a heavily trailered parody of Stanley Kubrick's *2001: A Space Odyssey* opening "Dawn of Man" scene. In *2001*, we see the Sun rise above the primeval plains of Earth, to the rising soundtrack of Richard Strauss's Nietzsche-inspired tone poem, *Also Sprach Zarathustra*. A small band of human apes are on the long, pathetic road to racial extinction. In *Barbie*, it's a small gaggle of little girls that populate the deserted plains, and we hear Helen Mirren deliver a cinematic voice-over, "Since the beginning of time, since the first little girl ever existed, there have been . . . dolls." Then, as Strauss plays, Mirren's voice-over continues, "But the dolls were always and forever *baby* dolls . . . until . . . ." Like the apes in Kubrick's movie, the girls now look up to see, not the mysterious monolith of *2001*, but Barbie, monumental, standing tall and colossal, like Antony Gormley's *Angel of the North*, cast not in steel but plastic.

In Kubrick's picture, the human journey begins with the epiphanic sight of the monolith. One of the hominids proudly hurls an animal bone into the air and, in an astounding cinematic ellipsis, the bone instantly morphs into an orbiting satellite, and three million years of human evolution is written off in one frame of film. In *Barbie*, the female journey begins with the revelation of Barbie. One of the girl gaggle is so inspired that she wields an old-school baby doll by the legs and smashes it against another baby doll, throwing the doll into the air where it instantly morphs into a huge, pink, sparkling bubblegum logo of Barbie. After millions of years of evolution, Barbieland is here.

And what a world it is. The movie follows a day in the life of a stereotypical Barbie. This is the "main" Barbie, the "Barbie you think of when someone says 'think of a Barbie.'" Stereotypical Barbie lives in Barbieland. Let's park that fact for a moment, before we describe Barbieland, and take a minor detour into the created literary world of English novelist, Jasper Fforde. Fforde's BookWorld is a fictitious and complicated place that acts as a "behind-the-scenes" area of books place where fictional and literary characters actually live their lives between active readings of their particular book by readers in the real world. The "engine room" of BookWorld allows the reader to read books using a complex system to supposedly continue the images being created in the reader's mind.

Barbieland is a similar realm where the created characters of literal Barbies live and interact while their doll bodies are being played with in our (real) world. The true selves of Barbies live here, though they take on the traits of whatever's happening to their doll bodies in "real" play. The Barbie dream houses don't have walls, just like in life. The Barbie world doesn't come with food, just adhesive decals and plastic pieces. Thus, Barbieland is kind of magic; outfit changes occur spontaneously, depending on play activity, and Barbies float around from one level of their dream houses to another.

With a superb parodying of superhero exceptionalism, Barbieland is a soundstage utopia of ticky-tacky box houses painted in hyper-saturated color, a pastel-pink Shangri-La. What makes this place so particularly special? Why, "thanks to Barbie, all problems of feminism and equal rights have been solved." Barbieland is a fantastic world in which big-haired dolls can be whatever they choose (physicists, presidents, lawyers, doctors, etc.). And the sheer perfection of the

Barbieland exemplar inspires copycat feminine achievement in the "real world;" ("we fixed everything so all women in the real world are happy and powerful!") The Barbieland that Gerwig has conjured up is a little like the Matrix, in which, as in that exhibition at the Louvre, there are multiple models of Barbie, rather than copious clones of Agent Smith. All the women here are Barbie.

And at the locus of all these parodic trifles, Stereotypical Barbie is a role so perfectly played by Margot Robbie that when narrator Helen Mirren makes a snarky joke about the casting, the audience doesn't even mind. But wait, there is trouble in paradise. In this realm of rictus smiles and endless sunshine, Stereotypical Barbie suddenly becomes haunted by anxious thoughts of sadness and death: Is death the aim and purpose of life? Is it true that to live is to suffer? Can it be that the triumph of death is inevitable, and that existence is a constant dying? If death is the first head of Cerberus in Barbie's nightmare, the other two heads of the Barbie apocalypse are, heaven forbid, the flat feet and the cellulite she develops. Shock, horror; Barbie has become a real woman. What possible way is there out of this crisis?

THE PHYSICS OF BARBIELAND

Stereotypical Barbie visits Weird Barbie ("she was played with *too hard*") and finds that a wormhole has opened between this world and the next. Just as Kubrick's *Odyssey* is a journey for "mankind," Barbie's odyssey is a ride for womankind into *our* reality. Accompanying her on this journey is Stowaway Ken (Ryan Gosling delivering a knockout comic performance), and the pink-land pair promptly discover the patriarchy, in which men (and horses) rule the roost.

Indeed, the "physics" of the real world is very different to the physics of Barbieland. The realm of Barbieland is governed not by the normal laws of nature, but by the childlike whimsy of playtime: Barbie drinks from an empty cup, rebounds back off plastic water, and simply floats down from her top-floor bedroom to her parked car below, as if carried by the guiding hand of an invisible girl. The dialogue in Barbie is also a fascination. Kubrick's *Odyssey* is, in part, a portrayal of an insipid future dominated by corporations and technology. The banality and vacuity of human characters is sharply and ironically contrasted by the robust intelligence of the infamous computer HAL. In *Barbie*, the happy chatter of the rather bare-bones dialogue is a little like the inner voice of a nine-year-old's creative imagination. Every day declared forever and ever to be the best day ever, and every friend the very best of best friends.

And so, when Barbie and Ken get to the real world, they're stunned to find that our reality bears little or no resemblance to their estrogen-driven utopia the other side of the wormhole. After all, the physics of feminist Barbieland boasts all-female government officials, trash collectors, and Nobel laureates, along with a pack of devoted and pliable Kens, forever grateful to bask in the reflected glory of mass Barbie-dom. Barbie and Ken react differently to the physics of our reality. Barbie is dumbfounded to discover she isn't the all-conquering super-role-model she imagined herself to be. At Mattel HQ, Will Ferrell stars as the doodle-dasher of childhood dreams, insisting that Barbie gets back in her box. Worse, Barbie is told by Goth-teen Sasha, "you've been making women feel *bad* about themselves since you were invented," declaring "You set the feminist movement back fifty years, you fascist!" In short, rather than saving the world, Barbie appears to have helped fashion a dystopia where "everyone *hates* women!"

SUPER-KEN

In the real world, Ken is encouraged to find the hegemony of a power structure that places him and his male kin on top. Galvanized by the new ideology of this reality, Ken carries the "good" news back to Barbieland, and before you can say "Mary Wollstonecraft," Ken has founded a full-blown patriarchy. Now, all the previously empowered Barbies are brainwashed into compliant, glorified barmaids, mere slavish stewards to male pleasure. The challenge for Barbie becomes clear as crystal: she must open the eyes of her sisterhood to the potential that there's more to womanhood than merely being an accessory to a man or a flawless exemplar of femininity.

The *Barbie* movie is naturally a female triumph, directed by Gerwig, a woman included in the annual *Time* 100 list of the most influential people in the world in 2018, and coproduced by Margot Robbie, ranked as one of the world's highest-paid actresses by *Forbes* in 2019, yet the film's production was still savvy enough to let Ken not only be the butt of many of the jokes, but also be self-aware, self-referential, and pretty damn complex. Little wonder many cinema critics suggested Ken might just qualify as the greatest role of Gosling's career to date.

Gosling's role is an unexpectedly challenging one. His character is a jewel set in a firmament of a film. A complex movie which adopts a variety of inquiries: the doll icon is explored through cinematic conventions as varied as the faux documentary, the movie musical, and the traditional hero's odyssey, like *2001*. Thus, whereas Ken could have merely been a one-gag guy, the production has gifted him a complex inner turmoil. Ken wants desperately to be "seen" by Barbie, but his neediness morphs into a toxic masculinity that infects the other Kens.

His reckoning is also pleasingly profound. The movie's high point arguably arrives when Barbie declares that she wants to be a real girl, yet most of the emotional heft came ten minutes before, when Ken is seen wrestling with the idea of internal validation, and his epiphany comes when he realizes he can never be happy until he understands who he is outside of his relationship with Barbie. Gosling plays the arc with great comic effect, but also with a tasty dash of palpable rawness.

SUPER-BARBIE

Barbie is a creative and hugely imaginative thought experiment. It's not just about the Barbie doll. Not even simply about her complex legacy and what she represents. It's also about what it means to be a woman and, by reflection, what it means to be a man. *Barbie*'s field of discourse is an alternate reality where thoughts and emotions are explored through song, music, and dance, along with much that would seem magical in real life.

But Barbieland is also a paradise of female empowerment, a realm diverse in representations of female distinction and refinement. The movie makes clear that Barbie has taken on multiple meanings and identities since her bathing-suited emergence in 1959. The Barbie concept is, in a sense, *all* women, and a reminder that women can do anything. The patriarchy, and male supremacy, for all its ubiquity, is a relatively recent abnormality. Evidence suggests that patriarchal societies date back barely ten thousand years. And humans likely evolved as an egalitarian species, staying that way for hundreds of thousands of years. One piece of evidence for this equality is in the similar size of human females and males. We humans exhibit the least size disparity of all the apes, suggesting male dominance is not the driving force for us. Indeed, sexual equality in our early ancestry would have been evolutionarily advantageous. "Parents" who were emotionally invested in both girls and boys gifted our ancestors a survival benefit. The investment meant a fostering of vitally broader social networks that we depended on to exchange resources, genes, and cultural knowledge.

Genetically speaking, and during early development, the gonads of the human fetus remain undifferentiated. In other words, all fetal genitalia are the same and are phenotypically female. It's only after about six or seven weeks of gestation that the expression of a gene on the Y chromosome induces changes that result in the development of the testes. In Barbieland, the Barbies (accomplished, happy, and beautiful as they are in their various jobs and roles) run a supportive and productive society. The Kens have no jobs or purpose, but exist merely for the Barbies. That is, Ken lives for Barbie, longs to unite, pines for her love. As Greta Gerwig has said in interviews, it is Barbie, not Ken, that's the main draw of Mattel products. And the implications of that are fascinating, "Ken was invented after Barbie, to burnish Barbie's position in our eyes and in the world. That kind of creation myth is the opposite of the creation myth in Genesis." Not so much Eve serves Adam as Ken serves Barbie.

The main theme of the *Barbie* movie is the envious women-hating current that still lurks deep in the male outlook. Even the devoted Ken is secretly something of an incel. Meanwhile, through Barbie herself, we watch as all the achievements and dreams of women are deconstructed and destroyed by the kind of men who can only feel powerful themselves through their control of powerful women.

Witness what happens when Barbie realizes her utopian dream-world is a disaster. Barbie develops melancholia. She starts to doubt herself. She feels plain and ugly. Insignificant. A failure. She hates herself. Greta Gerwig's screenplay is part inspired by Mary Pipher's 1994 nonfiction book, *Reviving Ophelia: Saving the Selves of Adolescent Girls,* about the sudden, mass depression and confidence crisis that hits girls around puberty. Pipher's contention is that, while the women's movement has focused on the empowerment of adult women, teenage girls have, to some extent, not only been neglected, but also need intensive support due to their undeveloped maturity.

As Gerwig explained to *Vogue* about the effect on young females, "They're funny and brash and confident, and then they just, stop . . . All of a sudden [girls think], *Oh, I'm not good enough.*" In addressing such issues, *Barbie* is both ambitious and thoughtful, which is quite the feat given the movie's main character is essentially (admittedly superhero) plastic. Barbie may be plastic, but *Barbie* is profound.

CHAPTER 7

IS STORM WISE TO TINKER WITH THE WEATHER?

(. . . in which we mull over the consequences of messing with chaos theory)

Exterior. Wooded hillside. A sudden flash of light. The wind blows so violently now that Sabretooth nearly misses two figures standing only a few yards away, mere silhouettes in the icy haze. A closer look tells us it is a man and a woman, they wear strange uniforms of form-fitting material, the woman's face is bare, revealing dark skin and penetrating eyes. Storm looks down . . . concentrating her intense gaze. The wind whips further, and her skin dimples with goosebumps as the temperature drops and the water . . . begins to freeze. Storm then further forces the temperature down, freezing the ice thick and thicker . . . The snow and wind are now violently raging.

—Ed Solomon and Chris McQuarrie, *X-Men* early screenplay draft (1999)

It used to be thought that the events that changed the world were things like big bombs, maniac politicians, huge earthquakes, or vast population movements, but it has now been realized that this is a very old-fashioned view held by people totally out of touch with modern thought. The things that really change the world, according to Chaos theory, are the tiny things. A butterfly flaps its wings in the Amazonian jungle, and subsequently a storm ravages half of Europe.

—Neil Gaiman, *Good Omens: The Nice and Accurate Prophecies of Agnes Nutter, Witch* (2006)

You've never heard of Chaos theory? Nonlinear equations? Strange attractors? Living systems are never in equilibrium. They are inherently unstable. They may seem stable, but they're not. Everything is moving and changing. In a sense, everything is on the edge of collapse . . . God creates dinosaurs. God destroys dinosaurs. God creates man. Man destroys God. Man creates dinosaurs. Dinosaurs eat man. Woman inherits the Earth.

—Michael Crichton, *Jurassic Park* (1990)

EXPLOITING NATURE

Ever since the first irresistible rise of science, the mission was not just to explore nature, but to exploit it. The machines of science were created with nature's dominion in mind, and science fiction took that mission into a far-flung and imagined future. The explorer who seeks to penetrate space, such as in Jules Verne's *Journey to the Center of the Earth*, wishes to possess nature absolutely for science. To reach the core of the world is to achieve completion, to pierce the living heart of nature, the glittering prize. Likewise, in H. G. Wells's *The Time*

Machine, where the time traveler sets out to navigate and dominate time. His doom-laden discovery for science and for the human race? Time is lord of all. The significance of the story's title becomes clear: humans are trapped by the mechanism of time and bound by a history that leads to our inevitable extinction.

And high on the list of the challenges facing science in its quest to tame nature is the weather. To control the chance and capricious whim of forces that can't even be seen, let alone predicted. Most people find that the challenge of forecasting the weather is that it's right too often for us to ignore it, and wrong too often for us to rely on it. Weather often seems to be one of nature's greatest secrets in its mood of mockery of humans. And yet the abiding rain has soft sculpting hands with the power to cut stones and chisel sheer drama into the shapes of the very mountains while creating rainbows in an apparent apology for angry skies. Little wonder that scholars have long dreamt of controlling the weather. The right amount of rain and sun means healthy crops, shelter, and prosperity; too much or too little, hunger and death.

READ THE SKY, DIVINE THE WEATHER

The dream of weather control is ancient. Early farmers learned to make a clock and a calendar of their changing sky and consulted almanacs for astronomical guidance in deciding when to plant and harvest their crops.

The hunter-gatherers who preceded farmers also used the sky as a weather calendar. As a Cahuilla Indian in California told an academic in the 1920s:

The old men used to study the stars very carefully and in this way could tell when each season began. They would meet in the ceremonial house and argue about the time certain stars would appear, and would often gamble about it. This was a very important matter, for upon the appearance of certain stars depended the season of the crops. After several nights of careful watching, when a certain star finally appeared, the old men would rush out, cry and shout, and often dance. In the spring, this gaiety was especially pronounced, for . . . they could now find certain plants in the mountains. They never went to the mountains until they saw a certain star, for they knew they would not find food there previously.

Humans even used reckoning machines to help them in their reading of the sky and the timing of the seasons with their associated climates. Stonehenge was one such reckoning machine. Built in the Stone Age, it was an ancient time-reckoning machine whose moving parts were all in the sky. The Great Pyramid of Giza was another. It was aligned with the pole star so that the ancient Egyptians could read the seasons from the position of the pyramid's shadow. And the stone medicine wheels of the Plains Indians of North America were another reckoning machine. The wheels were used to tick off the rising points of brighter stars, thereby announcing to the wheel's nomadic architects the arrival of the date to migrate to seasonal grazing lands. Moreover, it has been argued that the twenty-eight poles of Cheyenne and Sioux medicine lodges were used to mark the days of the lunar month. In the words of Black Elk, a priest of the Oglala Sioux, "In setting up the sun dance lodge we are really making the Universe in a likeness."

Thus, given the long human history with the weather, the twentieth-century creation of a fictional superhero who could master the weather was somewhat inevitable. And so Storm, who was born Ororo Munroe, first appeared in 1975 and is most commonly associated with the *X-Men*. Storm's backstory is a tale of hereditary mutation. Her mother was a Kenyan princess from a long line of African witch-priestesses with signature blue eyes and white hair. The witch lineage is a line of mutants born with the superhuman ability to control the weather. Very handy. On the untimely death of her parents, Storm ends up being worshiped as a goddess once her powers manifest and is eventually recruited by Professor X. Storm soon becomes a member of the X-Men, leading them from time to time, while also working with the Avengers and the Fantastic Four.

WHETHER WIELDING THE WEATHER IS WISE

Yet, might Storm's weather wielding be rather unwise? Given our current problems with climate change, is meddling with the weather to be admired? For example, what about the so-called Butterfly Effect? This idea first found its voice in science fiction with Ray Bradbury's moral fable, *A Sound of Thunder*, written in 1952. In Bradbury's tale, a time-tourist wreaks temporal havoc by treading on a prehistoric butterfly and unleashing an alternate timeline. The story told of sensitive dependence upon initial conditions. It was written a full ten years before an early pioneer of chaos theory, Edward Lorenz, developed its principles for the scientific community through mathematics and meteorology. Since then, writers like Neil Gaiman have written about how "a butterfly flaps its wings in the Amazonian jungle, and subsequently a storm ravages half of Europe" and Michael Crichton in *Jurassic Park* tell the dramatic tale of what happens when living systems, inherently unstable and forever moving and changing, are pushed over the edge and collapse.

Let's conjure up our own Storm weather-wielding scenario and work through the possible consequences of her actions. For the sake of argument, let's say Storm is battling with Magneto. Hardly a stretch of the imagination for, as good and bad guys go, Magneto rarely seems to know what side he's really on. And since he's one of the most powerful and dangerous members of the Brotherhood of Evil Mutants, he's more than a match for Storm with his forceful ways, manipulation of electromagnetism, and sheer will to do something insane—like shut down the internet or some other despicable atrocity.

STORM VERSUS MAGNETO

The battle doesn't start well for Storm, and Magneto quickly gets the upper hand, so Storm comes to an epic conclusion: fight fire with fire, or at least electromagnetism with electromagnetism. She plans to focus all her spooky Kenyan witch power and aims to telepath the sum of all lightning strikes currently occurring around the globe. Her noble goal is to zap Magneto's ass once and for all. Storm knows there's power in a lightning strike and, as lightning is electricity, it should have a decent zap factor, as long as she can get all the lightning to strike where Magneto stands.

Storm sets her witch-priestess thoughts in train. She knows a typical lightning strike harbors enough energy to power a typical home for a couple of days. She also knows that even in places on the Earth's surface that see a lot of lightning such as the Congo, the power delivered to the ground by lightning is a million times less than the power brought by sunlight. But Storm's master plan is to conjure up all the lightning. To make a mega-bolt, in which all the lightning comes down in parallel, bunched up in one big bolt.

Storm is aware of the fact that the main channel of a lightning bolt, the zap that carries the current, is only around one centimeter across, so Storm's mega-bolt, the sum of all strikes and containing around a million discrete bolts, will be more than enough to zap Magneto where he hovers. Storm thinks about the way that Magneto is often boasting about his powers in measures of Hiroshimas: "Wow, did you see that, Storm? Niagara Falls may deliver the power of a Hiroshima every eight hours, but I just did it in a single second," or "the soft breeze that blows across a prairie might carry the kinetic energy of a Hiroshima, but, Storm, I just made the hair stand on end of everyone in midtown Manhattan." That kind of thing. But now, Storm is about to get her own back. The mega-bolt coming Magneto's way will zap him with two atom bombs' worth of energy.

Somehow (this is superhero stuff, after all), Magneto divines Storm's intent. He quickly does some mental calculations of his own. These are mostly focused on the ideas of: plan A, getting out of the way, or plan B, somehow dissipating the power that is about to fall from heaven above. Totally out of character, Magneto doesn't think of the incoming energy in terms of Hiroshimas; maybe he's in denial, or simply doesn't want to credit Storm with such superhero proportions. Instead, he figures there's enough electrical energy in the imminent mega-bolt to power a games console and plasma television for many millions of years, or to feed America's hungry domestic need for electricity for just five minutes.

Magneto mulls through his options. The girth of the mega-bolt would measure about the same as the center-circle of a basketball court, but the damage it would do would wreck the very court itself. Not much room for Magneto to maneuver. Besides, inside the mega-bolt, the air would be transformed into high-energy plasma, and that means the light and heat from the bolt would extemporaneously ignite surfaces for miles in all directions. The very shock wave of the mega-bolt would fell trees and flatten buildings, leaving Magneto with a dwindling list of choices.

Instead, he considers plan B: Rather than merely sidestepping the mega-bolt, somehow deflect it elsewhere. It would easily be within Magneto's powers to draw some kind of lightning rod toward him and try to ward off the mega-bolt that way. But it's not the best of his options. For one thing, no one really knows how lightning rods work.

One theory is that they ward off bolt power by "earthing" charge from the ground into the atmosphere, thereby lowering the difference in electrical potential between the clouds and the ground. This is thought to reduce the probability of a strike. But, since Storm is doing this on purpose, a rod is hardly likely to help Magneto in his current predicament. Say, for example, that Magneto was able to use his powers of affinity to summon a copper cable just touching the ground. Sure, the brief torrent of current from the mega-bolt would be conducted by the copper without melting it. But when the bolt shot down to the bottom of the rod and made contact with the ground, there would be an explosion of molten rock to deal with, and Magneto was less happy with lava than he was with lightning.

But then the coin drops in Magneto's mind. He remembers Storm is deeply claustrophobic. Magneto recalls from his knowledge of *X-Men* history that, although she is a very strong woman, Storm, like Superman, has one main weakness (in her case, claustrophobia). The phobia dates back to her childhood in Cairo, when a jet crashed into her home, killing her parents. The impact brought the house down on young Ororo, burying her alongside her dead parents under the rubble, having to wait days for rescue, lying next to their corpses. The incident left Storm with a severe phobia of enclosed spaces, a phobia that Magneto could now exploit. Not for nothing is he a member of the Brotherhood of Evil Mutants. Having so much experience with lifting objects such as missiles, guns, cars, and even the entire Golden Gate bridge, Magneto is easily able to create a made-to-measure metal cage in which to fence and befuddle the witch-priestess.

That's the thing with the nonlinear equations of chaos theory. Storm messes with Magneto. Storm wields the weather. Magneto messes with metal. Storm catches a new case of claustrophobia. You can cage the wielder, but not the weather.

CHAPTER 8

WHAT OF WITCHES, WEDNESDAY ADDAMS, AND WANDAVISION?

(. . . the story of the cinematic witch: from the origins and a potted history of witchcraft, through Willow and Wednesday to the wonder of WandaVision)

[W]e conceive the Devil as a necessary part of a respectable view of cosmology. Ours is a divided empire in which certain ideas and emotions and actions are of God, and their opposites are of Lucifer. It is as impossible for most men to conceive of a morality without sin as of an Earth without "sky." Since 1692 a great but superficial change has wiped out God's beard and the Devil's horns, but the world is still gripped between two diametrically opposed absolutes. The concept of unity, in which positive and negative are attributes of the same force, in which good and evil are relative, ever-changing, and always joined to the same phenomenon—such a concept is still reserved to the physical sciences and to the few who have grasped the history of ideas.

—Arthur Miller, *The Crucible: A Play in Four Acts* (1953)

NEVERMORE IS A MAGICAL PLACE

Late in 2022, a supernatural female became a smash hit. An "outcast" with a witch as an ancestor, Wednesday Addams was the star. And, just three weeks after its release, that show, the young-adult series *Wednesday*, the Netflix Addams Family reboot, directed in part by Tim Burton, racked up a flabbergasting billion hours' worth of views, a Netflix benchmark previously passed only by *Stranger Things* and *Squid Game*. *Wednesday* is a wonderfully macabre remedy to white-picket-fence American schmaltz, steeped in good old-fashioned gothic convention and all the best bits of modern Halloween.

Thus quickly becoming *Netflix*'s second most-watched English-language series, the awards and nominations started to stream in. Two Golden Globe nominations, for *Best Television Series—Musical or Comedy* and *Best Actress—Television Series Musical or Comedy* for Jenna Ortega, who plays the titular character, were followed by Primetime Emmy nominations for *Outstanding Comedy Series* and *Outstanding Lead Actress in a Comedy Series*. The awards were accompanied by a social media frenzy. Myriad fans took Instagram selfies in their best Wednesday cosplay. Teeming throngs of TikTokers carefully recreated Wednesday's Siouxsie and the Banshees–inspired dance, in which Jenna Ortega's viral choreography to *Goo Goo Muck* seemed to single-handedly revive Gothic subculture.

Wednesday was a magic trick. Given that Netflix had a dismal year of plummeting stock and lost subscribers, its *Wednesday*-induced recovery was miraculous. *Wednesday* was a potpourri of fail-safe storylines, laced with a little something for everyone. And this young-adult dark comedy centered on a supernatural, coming-of-age murder mystery with Wednesday Addams, already the daughter of a familiar and beloved television family, playing a kind of Dana Scully/Nancy Drew crossover. Wednesday is not just an underdog misfit weirdo. She just also happens to be the official "It Girl" of the moment.

Wednesday is the smartest kid in her new school, *Nevermore*, a Hogwarts-like remote boarding school for outcast "fangs, furs, stoners, and scales." With a ready arsenal of barbed rejoinders and erudite quips, she is eminently quotable. In the very first episode, we see her expelled from her "normie" suburban high school for vengefully un-bagging piranhas on the high school water polo team. It's payback for her little brother's bullying, which she makes clear by telling the severely bitten polo-boys, "no one tortures my brother but me." She's the ideal student. A talented polyglot, cellist, novelist, and fencer, she is nonetheless "an outcast in a school of outcasts," a dedicated misanthrope who bridles at human contact. "Sartre said hell is other people," Wednesday tells her inevitable therapist. "He was my first crush."

In the third episode, *"Friend or Woe,"* Wednesday's heritage as a witch is highlighted. An illustration in an old tome leads Wednesday to an incredibly vivid vision of her ancestor Goody Addams. In a wood, Wednesday envisions Goody about to be executed by Joseph Crackstone, the town's founder who is intent on killing all outcasts.

Various villagers: "Witch! Repent! Begone! Witch!"

Joseph Crackstone: "Goody Addams! You have been judged before God and found guilty. You are a witch, a sorceress, Lucifer's mistress herself. For your sins, you will burn this night, and suffer the flames of eternal hellfire."

But, alone among the outcasts, Goody is able to escape.

Wednesday describes herself as single-minded, stubborn, and obsessive. "All traits of great writers, yes, and serial killers," as she says to Thing, the famous Addams Family disembodied hand. She has all the skills of a typical fictional witch: she's smart, morose, condescending, lethally blessed with alluring good looks but burdened with the baggage of past trauma (in her case the trauma is the passing of her pet scorpion, who was murdered when she was six; she's since vowed to never cry again). And, like the witches of the past, Wednesday is an outcast, even among the outcasts. Before looking at some other famous media witches, let's take a look at the evolution of the idea of witches like Goody Addams.

A BRIEF HISTORY OF WITCHES

The long history of witchcraft is complicated and can lead to more questions than it actually answers. Here are just a few questions you may have wondered about: Where on Earth did witches first come from? Why are they pictured arriving on broomsticks? And what exactly do we mean by the word "witch"? To help answer such questions, it'll really help to take a look into the minds of ordinary people and scholars in medieval England. By using old England as an example, we will gain a better grasp of how the idea of the witch evolved.

It's worth bearing in mind that the medieval period was a far cry from today's (mostly) rational world. In 1458, a pig was hanged for murder in Burgundy. In 1602, a French judge named Henri Boguet described an apple as being possessed by demons. And, a few years later, Italian Jesuits tried to calculate the physical dimensions of hell itself. So, when we take the following witchy journey, from the pre-Christian world to the burial mounds of the medieval English landscape (a place where an underworld realm of elves, demons, and familiars was alive in the popular imagination), we can begin to better understand the "common sense" of the developing times. And, out of this murky evolution, we will find how witches became the subject of the chilling persecutions of the sixteenth and seventeenth centuries.

First, let's address the commonly asked question: How long ago in history does the belief in witches begin? Arguably, the majority of readers might believe that the notion of witches is a Christian creation. But, actually, the seductive idea of the witch who flies in under the cover of darkness and draws power from shadowy cosmic forces to work and wield her magic antecedes Christianity by centuries. For instance, Homer's *Odyssey* describes the enchantress and a minor goddess, Circe, as a witch who has vast knowledge of potions and herbs and who can use her magic wand or staff to transform her enemies into animals. And the ancient Greek historian, Plutarch, mentions witchcraft in his treatise, *On Superstition* (c. AD 100).

Moreover, references to witchcraft occur in ancient Roman law. There are many such citations of illicit magic in Roman statutes, some of which were inherited by the Christian world. Archaeological scholars have also discovered hundreds of ancient Greek curse tablets, which in the ancient Greek world were known as katares, or "curses that bind tight." It appears they were invented by the Greeks themselves, with a considerable number centered on sporting or legal contests. The inscribed katares were placed in graves, wells, or fountains, where the dead were expected to work their magic.

THE EMERGENCE OF HERESY

Many centuries later, deep in the Age of Faith, the allegedly pious decided there was a problem with heresy. Why? Because across the world, the social vacuum left by the collapse of the slave-owning societies of the ancient world led to the rise of a feudal economy. It was an economy so fragmented and localized that it needed no radically new worldview. No paradigm was necessary to replace the ancient one that had served the classical world.

The victory of Christianity in the West meant that, from the fourth century on, intellectual life was confined to church men, and during the early Middle Ages, the history of thought over the lands of the disappearing Roman Empire was the history of Christian dogma. The church fathers had set about their mission impossible, to integrate the more innocuous elements of the ancient wisdom into Christianity. Much of the old philosophy had already found its way in, by stealth. But the Old Testament and classical culture were unhappy bedfellows. As some of the more progressive philosophers tried to crowbar in some safer aspects of philosophy, controversy was inevitable, and beginning in the fourth and fifth centuries, great disputes and heresies raged.

THE EMERGENCE OF THE SHAMAN

Constant attempts were made to defeat heresy, and this campaign brought to light a number of figures who were pretty tricky to square with Christianity. Such figures were usually reported without mention of witchcraft, but nonetheless later led to the creation of the idea of the heretic witch. For example, consider a female figure known as Perchta (or Bertha or Befuna), a goddess in Alpine paganism in the Upper German and Austrian regions of the western Alps. Perchta was the female embodiment of winter. She punished disobedience and rewarded goodness. She was most often depicted as an old hag, as she symbolized winter and the cold, so it didn't take long for scholars to note her likeness to the witches with whom they were familiar from classical literature.

Pagan figures like Perchta had a long history. The methods of the early shaman were based on mimicking, and sympathizing with, the workings of the cosmos. (Let's not forget that the word shamanism comes from the Manchu-Tungus word, *šaman*. The noun is formed from the verb ša, "to know;" so a shaman is literally "one who knows.") The shamans recorded in historical ethnographies have included women and transgender individuals, as well as men, and are of every age from middle childhood onward.

From evidence archeological scholars have amassed of the cave art of Western Europe, it seems these shamans were already established in the Old Stone Age. The cave paintings of the Trois-Frères in the Ariège department of southwestern France are a fine example. A painting at the Trois-Frères shows a shaman, or sorcerer, wearing stag's horns, an owl mask, wolf ears, the forelegs of a bear, and the tail of a horse. The value of such Animagus behavior may have been to ensure a successful hunt, essential to the tribe.

At first, the shamans would use likenesses, and later symbols, to perform an operation on something that would be considered transferable to the real world. An unbroken thread links these ancient symbols to those used with such success in modern science. Another feature of primitive thought, which at some point separated itself from imitative or symbolic magic, was the idea of the influence wrought upon the real world by spirits. The idea of a spirit probably emerged from the reluctance to accept the fact of death. Early spirits were very worldly, members of the tribe who had since passed on. But the idea evolved that it was necessary to win, or regain, the favor of a spirit, now god, by doing something that pleased them.

The old idea of spirits split into two very different forms. On the one hand it transformed into the idea of spirit as an all-powerful being, or god, that was to become the central figure in religion. And on the other hand, the spirit became divorced from human origin to become an invisible natural agent, such as the wind, or the assumed active force behind chemical and other crucial changes. This second idea of the spirit was to become hugely important in the evolution of the understanding of "spirits" and gases in science.

So, science and witchcraft are far more entangled than we might think. At first the rituals of shamanism would have involved most of the tribe. But, in time, cave art shows solitary figures of the tribal Animagus, dressed as an animal, who appears to have some special place. In primitive tribes today, there still exists medicine persons, or magicians. They are held in high esteem, as they are thought to have a peculiar relationship with the forces of nature and the cosmos. To some extent, they are set apart from the normal work of the tribe. And, in return, they exercise their shamanistic arts for the tribal good. They are keepers of learning and knowledge. They are the forerunner, the lineal cultural ancestor, of philosophers and scientists. And these are the kinds of figures and individuals that the church declared to be heretics. Witchcraft to the ignorant, simple science to the learned.

THE EMERGENCE OF THE WITCH: A TIME TRAVEL TALE

So, gradually, the idea of the witch developed. And a broad common consensus occurred, namely that a witch is someone who uses magical entities, such as the powers she holds within herself, to potentially harm others. A witch didn't have to be female. And they didn't have to have a black pointy hat and a broomstick. But they did have to be outcast in some way. Disfigured, maybe. Or crooked. They had to be like the dead in some way, such as hard, or infertile. And they had to hate. Again, like the dead, who hate the living, the witch also "hated." There is no specific time when this common idea was born.

Let's take an AI-guide-assisted time travel trip back to the early modern period of British history, say around four hundred years ago. The time travel pod has dropped you onto the side of a gentle rolling hill. As soon as you get out of the travel pod, you notice that, below your feet, there are curious looking lumps in the grass. As you wonder what they are, the AI guide tells you through your earpiece that they are prehistoric burial mounds.

We spoke earlier of the "common sense" of the medieval mind. You know, hanging a pig for murder, describing an apple as being possessed by demons, and trying to calculate the physical dimensions of hell; *that* kind of thing. Well, as you stand on the hillside, above the prehistoric burial mounds, imagine now that the clever AI-guide is not only able to convince you that you are a member of the community that flourished here centuries ago, but also able to somehow make you "feel" the "common sense" of the times. This "feel" for the old times would enable you to envision how some members of your village wander here often. Now you wonder: Are they merely looking for herbs to pop into their porridge, or are they wandering for other, darker reasons?

Your imagination is now visited by a vision that, among the young girls in the village, it is whispered that if you wander to this place at midnight on All Hallows Eve, you pay witness to the dead not only rising up but also riding along the rocky road to the market cross. As they are not able to pass the cross, they stop there. It's diabolical luck to cast your eyes upon them, but if you catch the eye of one of the sinister riders, you might be gifted supernatural powers of healing and prophecy from which you could make your fortune.

What really lies in the Earth below your feet, beneath the humps of stone? The "common sense" of the times suggests some kind of realm of the dead. The church man in the village says that the dead who remain in the ground are those condemned to hell. Others say that the dead riders are actually wreathed in flames, and that their saddles are made of searing-hot iron. Those same people also say that if you do manage to get any supernatural power from the riders, it's the powers of devilry and hell.

And yet other, older people in the village have another belief. These elders believe that the dead who stay in the Earth are not demons. They are elves. They believe that, beneath the curious looking lumps in the grass is a lush, though sunless, land where the elves feast and sing and dance and make merry for their favorite mortals. But the elves remain dangerous, particularly if crossed. The dead pine for the lives they lost. And that means they want to steal it back from the living. They stay where they were buried. The cleaving of body and soul may take many months or even years, yet may never happen for those destined for damnation, so the sites where pagans buried their dead are particularly fraught. The places of the pagan dead are like a site of toxic waste. They may be buried, but by no means does that mean they're gone for good. Anyone sinister enough to feed them, especially on blood, can possibly put them to mischief. And, in this strange and unsettling set of beliefs, is the origin of the familiar of the witch, which now begins to take form: blood-fed like the dead and, also like the dead, malevolent.

And so the idea of the witch's familiar developed. Most often a small animal, but sometimes even as tiny as a housefly, the familiar was fed by the witch and, in return, it might reluctantly act out her will. The provenance of the idea of the familiar was, actually, a type of fairy known as a hob, or household brownie. Such creatures were said to hanker after cream, and so they could be placated by constant offerings of it, else they might make like poltergeists. Thus, it was thought that familiars could be put to mischief, ruining the work of other householders.

Incidentally, the idea of familiars may also be associated with the Norse fetch, or fylgia. This is a person's double, an entity that can also shape-shift to animal form. The fylgia is tied to a person's fortune or luck. But to the antiquated seventeenth-century church, the familiar was just a devil. There is reference to familiars in one of Britain's famous witchcraft trials. The Chelmsford witch trials of 1566 mention a familiar which resembles a human being. The notion that you can detach a part of yourself, a part that could look just like you, and dispatch that part to work your will on others, is a dominant idea in witchcraft.

But, whatever their provenance, the church's ideas of familiars were appropriated from popular pagan underworld of ideas and tales. It is said that, in some peculiar places in old Britain, common folks could "feel" the underworld beneath the soil: the weight of the past and the freight of its dead.

People living at that time were so superstitious and under the heel of church men, if they suspected one of the neighbors to be a witch, they were encouraged to never let her have the last word in a conversation. Anything the witch utters must needs be thrown back at her, before her utterance infected you. Neither should you let a witch give you anything, especially things associated with food, and certainly not food itself. You were never to let a witch cross the threshold of your house. Witch marks were to be used to stop her from entering your house, and to stop her familiars crossing the threshold into your home. Such witch marks were ancient boundary spells. A common traditional mark was a spiral in which the roaming entity will get lost. In the early modern age of the church, a common adopted boundary spell was the interwoven initial M, for the Virgin Mary.

THE MEDIEVAL CHURCH AND THE WITCH

We've looked at shamanism, heresy, and the "common sense" beliefs of ordinary people. Let's focus now on how the medieval church viewed witchcraft and how the church managed to shift the culture toward the persecution of women. It was in the eleventh century that *scholasticism* became dominant. This was the meeting of church and ancient Greek philosophy that we mentioned earlier. More specifically, scholastic philosophy was the system of theology and philosophy, taught in the medieval universities of Europe, which was based on the logic of Aristotle and the writings of Christian Fathers that emphasized tradition and dogma.

Under the auspices of scholasticism, all of God-created nature became an object of scrutiny. The scholastics would then try to create a paradigm which applied to all things. In this context, the inquisitorial eye of the scholastics started to fix itself on those parts of pagan folklore which had previously either been conveniently tucked away out of sight or, if possible, absorbed into Christian dogma in earlier times.

By the 1590s, the last decade of Elizabeth I's, the 1590s, the concept of the witch in England had hardened into the form of an old and poor woman, blind in one eye or maybe lame, and prone to losing her temper over personal trifles. Her dry, twisted, and moribund body was toxic, and she was thought to be able to harm people and other creatures just by talking or looking at them. Moreover, in the meantime, English law had been "refined" to reflect the research of European demonologists. These "scholars" ditched the idea that old women could conjure magic that defied the will of an omnipotent God and decided that it must be Satan who had all the power. Thus, the witch's familiar became a demon.

WITCH TRIALS

As a woman accused of witchcraft, your fate was very much in the hands of men. If accused, you would then be tried in one of three courts, either at a church court, at quarter sessions (which are local courts), or at the courts of assize (commonly known as the assizes), which were courts held in the main county towns and presided over by visiting judges from the higher courts, where you could be condemned to death. But the process was similar at whichever level of court was used.

Someone with a grudge could complain to the local justice of the peace (JP) that you had bewitched a child, say, or an animal, or perhaps a piece of food. Your ultimate fate would very much depend on the energies of the male JP, and how much of a witch-hunter he was. A fanatical JP might collate a considerable number of depositions (complaints from your fellow villagers), interrogate you, and extract a forced confession from you. Next, all this evidence is put before a jury of men, who mull over whether to take your case to trial. If tried, those villagers who submitted complaints take the stand and give their evidence under oath. You, as the accused witch, will also be asked to take the stand and your forced confession read aloud. In some circumstances, you may add to your confession, or deny whole or part of it, but that might just make you look inconsistent. Finally, the jury of men will decide on your guilt. *There is no counsel for the defense.* If found guilty, you could become one of the thirty thousand to sixty thousand women executed for witchcraft in the early modern era.

The medieval witch trials were a dark and complex period in history. While it's important to note that not all accused witches were women, the vast majority of those charged and persecuted during this time were, without doubt, female. There were several factors that contributed to the singling out of women during the medieval witch trials. Firstly, the trials occurred in a deeply patriarchal society, where women were generally considered inferior to men. This prevailing belief often led to the assumption that women were more susceptible to the influence of evil forces, making them more likely to be accused of witchcraft. Furthermore, the social roles and expectations placed on women during that time played a significant role. Women were expected to fulfill certain domestic and societal duties, and any deviation from these norms could be seen as suspicious or threatening. Traits such as independence, assertiveness, or knowledge in herbal medicine were sometimes misinterpreted and used as evidence against them.

And, as we have seen, another contributing factor was the prevailing religious beliefs of the time. Christianity played a crucial role in shaping the attitudes towards witchcraft, and many religious teachings associated women with original sin and temptation. This reinforced the idea that women were more susceptible to making deals with the devil or practicing malevolent magic. It's worth noting that the dynamics of the witch trials were complex, and various societal, cultural, and religious factors also influenced the persecution of both men and women during that period. But the singling out of women can be attributed to the combination of patriarchal attitudes, societal expectations, and religious beliefs prevalent at the time.

MAGICAL PICK OF THE WITCHES

Compared to the tortuous actual history of witches, since the advent of film and television, many of us have for some time got our stereotypes of witches from various media, particularly children's programs, which in turn, are often derived from children's books. That classic and deeply ingrained cliché of the old hag is still with us, but cinema and TV have depicted modern manifestations of the witch in such diverse ways that it would take up a whole book in itself. For example, Wicca influenced filmic portrayals of covens and the 1973 British folk horror film, *The Wicker Man,* is a fine example of this. Meanwhile, a fascination with African traditions fed into the 1966 British Hammer Horror film, *The Witches.* The alluring female witch has also played a prominent role in cinema and TV, from *Bewitched* of the 1960s to the contemporary witch genre of the likes of Wednesday Addams and Wanda Maximoff. As we've already considered Wednesday Addams, let's now briefly explore past screen portrayals of witches and look at how Wanda Maximoff fits into that history.

The popularity of witches like Wednesday Addams has been around since the start of film and TV, though it hardly rivals the abiding vogue for vampires, zombies, and demons. For too long, movie witch mythology came in the form of a decrepit old hag, cunningly disguised as a beautiful young woman. One thinks all the way back to Disney's first animated feature film, the 1937 classic, *Snow White and the Seven*

Dwarfs. In the movie, the Evil Queen character (also called the Wicked Queen or Queen Grimhilde, and based on the Evil Queen character from the 1812 German fairy tale *Snow White*) uses her dark magic powers to reverse engineer herself into an old woman, instead of merely taking a disguise as in the Brothers Grimm tale; this appearance of hers is commonly referred to as the Wicked Witch or the Old Hag.

Since those early days we've had a contrasting range of witches on our silver screens. From the innocuous Eglantine Price in *Bedknobs and Broomsticks* and Kiki in Studio Ghibli's *Kiki's Delivery Service* to the downright diabolical Minnie Castevet in Roman Polanski's *Rosemary's Baby* and Helena Markos, the alias for Mater Suspiriorum, who appears in both the original 1977 *Suspiria* and the 2018 movie remake and whose character is an old and wise witch who becomes more powerful as she amasses followers and sacrifices. There have been playful witches, such as Alexandra, Jane, and Sukie in *The Witches of Eastwick,* and highly underrated movie witches, such as Lady Van Tassel in *Sleepy Hollow*, and, of course, Hermione Granger in the *Harry Potter* franchise.

WILLOW ROSENBERG

Arguably, one of the most influential screen witches of the last generation or so was Willow Rosenberg from *Buffy the Vampire Slayer*. Originally a nerdy girl who not only contrasted Buffy's cheerleader vibe but also shared in the social isolation Buffy suffered after becoming a Slayer, as the seasons progressed, Willow evolved into a far more assertive and sensual figure. Realizing she was a lesbian, and becoming a powerful Wiccan, through everything Willow maintains her humanity and kindness to others. Mirroring the high school nerdiness of millions of *Buffy* fans, Willow's narrative arc expanded from mere slayer sidekick to chief supporting player to, eventually, a sixth season stint as full-on villain, when she began using her magical powers for deadly revenge after the death of her beloved, Tara. Brilliantly played by Alyson Hannigan, whose honest and endearing performance made Willow a consistent fan favorite whom, by the end, was just as important as Buffy when it came to saving the world.

From the Evil Queen to Willow, witches have been portrayed, for good or ill, as possessing differing strengths, an ability to cast spells, brew potions, and conjure circumstances to meet their ends. One of the most powerful recent superwomen witches in media is Wanda Maximoff. Not only can Wanda move things merely with her thoughts, she can also fix hallucinations in the minds of others, see their fears, and fiddle with the very fabric of reality. It's next level witchcraft.

BACKSTORY OF A SCARLET WITCH

Born in fictional Sokovia in 1989, as a child Wanda Maximoff was unaware that she was born a witch and unknowingly engaged in basic hex magic (a little like Harry Potter inadvertently speaking Parseltongue with a snake on a trip to the zoo when young). Along with her twin brother Pietro, and her parents Oleg and Iryna, Wanda lived in one of those cliched small and drab apartments that are commonly portrayed in western stories about eastern Europe. Naturally, as an escape from the unfathomable horrors of European culture, Wanda mysteriously enjoyed watching American sitcoms, as her father was in business selling knock-off DVD box sets so the family could practice English in the hopes of emigrating to that well-known utopia known as America.

However, utopia isn't always what it's cracked up to be, and Wanda's parents are killed by a Stark Industries missile strike on their apartment. For a couple of days Wanda and Pietro are trapped in the remaining debris of their dwelling, when another Stark Industries missile strikes.

But it doesn't trigger. Unknowingly, Wanda has cast a "probability hex" which has transformed the missile into a dud, though the (still unconvinced) twins remain in fear of an explosion until they are rescued. For the twins, the sheer trauma of the whole episode found its focus on the Stark Industries logo stenciled on the side of the missile. This left the twins with the heartfelt, and not unreasonable, belief that Stark himself was ultimately responsible for the deaths of Oleg and Iryna, as their passing occurred due to Stark profiting from sales of his company's weapons (again, not an unreasonable thought!).

Many years later, now young adults, the twins take part in political protests in their city (probably staged by the CIA) before volunteering for an "enhancement" program run by Hydra, the former science research division of Adolf Hitler's Nazi Party (clearly very bad). Hydra exposes the twins, and many other test subjects, to the Mind Stone, one of six fictional Infinity Stones in the Marvel Cinematic Universe (MCU). Elsewhere we learn that the Infinity Stones are the remnants of six singularities that existed before the Big Bang, apparently, which were compressed into Stones by cosmic entities after the Universe began and which were spread throughout the cosmos.

The Mind Stone can control minds, enhance the user's intelligence, and create new life. With respect to the first two features of the Mind Stone, during the Hydra program, Wanda sees herself as the Scarlet Witch. Wanda and Pietro become the only survivors of the enhancement and Wanda gains psychic abilities with her magic skills greatly enhanced. With regard to the third feature of the Mind Stone, namely creating new life, this is how Vision came into being. In *Avengers: Age of Ultron*, after being partially destroyed by Ultron, J.A.R.V.I.S. (Just A Rather Very Intelligent System), an artificial intelligence created by Tony Stark, is given physical form as Vision, using the Mind Stone. Ultimately, and without too much distracting detail, Wanda and Vision fall in love.

WANDAVISION

Bearing all this history in mind, along with so much more from the back catalog of the respective comic book characters, in early 2021 *WandaVision* was launched on Disney+. Based on the characters Wanda Maximoff/Scarlet Witch and Vision, *WandaVision* is the first TV series in the MCU produced by Marvel Studios. It shares a continuity with the films of the franchise, being set at a time after the events of 2019's *Avengers: Endgame*, and follows Wanda and Vision as they live their lives in the idyllic suburban town of WesTView, New Jersey. Idyllic, that is, until their reality starts evolving through different decades of sitcom homages and television tropes.

The premise is an amusing one. How do our romantic couple, she a telekinetic and reality-warping witch, he an enhanced super-android, settle down in white-picket-fence America, trying to live the lives of mere mortals? Based on this premise, *WandaVision* explores its answer through a series of loving homages to classic television sitcoms. The pitch is perfect. The scripts, the performance, the scene-setting, the lighting, and the cinematography. The initial episode is set in the 1950s, tapping into the likes of *The Dick Van Dyke Show* and *I Love Lucy*, which starred Lucille Ball and was the most watched show in the United States in four of its six seasons between 1951 and 1957. (As of 2011, episodes of *I Love Lucy* had been syndicated in dozens of languages across the globe and remained popular with an audience of forty million a year—the makers of *WandaVision* knew what they were doing!) Meanwhile, the second episode is based on *I Dream of Jeannie*, a 1960s sitcom about a sultry, two-thousand-year-old genie and the astronaut with whom she falls in love and ultimately marries. The 1960s vibe of *WandaVision* also leans heavily on *Bewitched*, which originally aired for eight seasons from 1964 to 1972 and is about a witch who marries a mere mortal, but vows to lead the life of an ordinary suburban housewife. *Bewitched* became the second-rated show in the

US in its debut season, remaining in the top ten for the first three seasons, and ranking eleventh for seasons four and five. Like *I Love Lucy*, *Bewitched* is still watched across the globe through syndication and on recorded media. After engaging with these titans of television history, *WandaVision* dips into the 1970s, *The Brady Bunch*, and beyond.

And yet all is not well in picket-fence televisual America. After all, when we last saw Vision in his cinematic guise, he seemed pretty much dead. And so we suspect that, from the get-go, and despite the fun and shenanigans of *WandaVision*'s hurry through sitcom history, all is not what it first appears in superhero-laced suburban paradise.

So, what possible witchcraft is at work here? We get a growing number of signs and signals that a deeper mystery and maybe malevolence is woven through the fabric of this "reality." At times, *WandaVision* is reminiscent of *The Truman Show*, the 1998 satirical sci-fi psychodrama which famously stars Jim Carrey as Truman, a man who grows up living a "normal" life that, unbeknownst to him, is actually played out on a large set populated by actors for a TV show in which Truman is the star. As journalist Erik Sofge said of *The Truman Show*, "Truman simply lives, and the show's popularity is its straightforward voyeurism. And, like *Big Brother*, *Survivor*, and every other reality show on the air, none of his environment is actually real." There's a reality behind the reality.

Cut to *WandaVision*. It's like a kind of cosmic superhero *Truman Show*, but drawn out over a dozen episodes, as the fabric of reality rips and Wanda does her darned best to sew up the rents that rear up. Whereas Truman plays detective on his daily life, Wanda tries to remake her reality whenever she spies disconcerting things. And yet, one can only stave off the Nazi Bee Man for so long, or ignore weird messages beamed by radio at the neighborhood planning committee, or disregard the birth of your twins a mere forty-eight hours after you apparently fall pregnant by your metal'n'Mind Stone partner.

Meanwhile, if that isn't enough to deal with, neighbors seem increasingly keen to dish the dirt of some secret truth while others, like Geraldine, with a "toughness and an ability to be a woman" in a male-dominated world, sporadically let slip secret facts such as Wanda had a twin brother who was killed by Ultron. The whole thing is stylishly, wittily, and deliciously done. The satire is fabulous fun, the gags are great, and the acting (particularly that of Olsen and Bettany, whose chemistry is so spot-on it would make Robert Bunsen blush), is first class. In short, *WandaVision* has the wonderful vibe of a show shaped by a production team who knew just what they were doing, where they wanted to end up, and were intent on enjoying the journey to that destination. Light and dark are witch-woven as one, the satire is never schtick, and all types of conventional TV tropes are added to the recipe to thicken the potion of a plot.

EVERYONE CAN MASTER A GRIEF, BUT SHE THAT HAS IT

Yet, the most welcome aspect of the show is that *WandaVision* is the first MCU production that has explored grief meaningfully with a female superhero from the movies. In the tradition of witches, Wanda has all the qualifying prerequisites for the trope of cinematic grief. She was orphaned young by a Stark missile. Her brother was killed in combat. And she was compelled to kill her lover, only to see time rewind and compel her to watch Thanos do it again moments later. As Shakespeare said in *Hamlet*, "When troubles come, they come not single spies, but in battalions." The same seems true for superheroes.

Wanda also experiences the alienation of a witch. Her isolation drives her to transmute her neighborhood and all its inhabitants into a kind of utopian suburbia, inspired by her most loved sitcoms. In this way she creates her own superhero *Truman Show*, set apart from the rest of the world so she can live happily ever after with Vision version 2.0 and the two imaginary kids she conjured into existence.

In short, Wanda finds consolation by becoming the villain. But such a situation can't last forever, as Wanda is meant to be the hero. And so, with each succeeding episode, the mask slips on her magicked reality, as people try to get Wanda to face up to her pain. A more villainous character among the cast compels Wanda to relive her worst moments. But another character, Monica Rambeau, offers solace and sympathy in grief, and tries to offer salvation for Wanda through empathy. Monica's olive branch is the signal to viewers that forgiveness is the way forward for the star of the show. Superheroes are human too. And so the real message of the series and the MCU is a profound one: grief has a profound effect on us, humans and even witchy super-humans. It may sometimes manifest in harmful or healthy ways, but it need not master us. Grief is governed by love as much as it is by sorrow. Or, as Vision so eloquently put it, "What is grief, if not love persevering?"

There have been other losses which have motivated superheroes. A main theme in *Avengers: Infinity War* is Thor's loss. The loss of his brother Loki, the loss of his home world of Asgard, and the loss of each and every member of his family. A main theme in *Captain America: The Winter Soldier* is Steve Rogers's loss of people and place. When he becomes essentially unstuck in time, Rogers mourns the time and place of the 1940s era from which he was unstuck, as well as the people he left there. In *Black Panther*, it is the loss of T'Challa's father that drives him to lead Wakanda against a would-be usurper. And the entire narrative of *Avengers: Endgame* is essentially the loss of half of the Universe and the survivor's guilt and grief felt by those left behind, which motivates them to right what wrongs they can.

That universal grief of *Avengers: Endgame* is then distilled and refocused into the way Wanda Maximoff processes her grief. And so it is that *WandaVision* is an aide-mémoire of how the most powerful humans experience emotions, hopefully, just as hard as we do. When superheroes choose to persist in the face of grief and adversity, when they celebrate love instead of pain, the tale told becomes the story of the ultimate heroic act. How very apt that such a story should center on a superwoman.

CHAPTER 9

ARE FEMALE SUPERHEROES LESS PROFITABLE?

*(. . . in which we lift the lid on superheroes
and corporate power)*

"Semiotics is the study of the use of symbolic communication. Semiotics can include signs, logos, gestures, and other linguistic and nonlinguistic communication methods. As a word, semiotics derives from the Greek sēmeiōtikós, which describes the action of interpreting signs."
— Whatis.com, *Semiotics* (2023)

"The astronomical growth in the wealth and cultural influence of multinational corporations over the last fifteen years can arguably be traced back to a single, seemingly innocuous idea developed by management theorists in the mid-1980s: that successful corporations must primarily produce brands, as opposed to products."
— Naomi Klein, *No Logo* (1999)

ICONIC SYMBOLS

The realm of the superhero is replete with unforgettable and iconic symbols. The bright red and yellow "S" logo of the Man of Steel. The concentrically circular star-spangled shield of Captain America. The uncompromising pirate-style skull of the Punisher. Each symbol characterizes the superhero's ideals and values. And these symbols and logos have become an indelible part of our popular culture, recognizable around the globe, symbols of bravery and strength in the face of adversity. Over the years, many consumer polls have been carried out, testing the zeitgeist of opinion on the latest top ten most iconic superhero symbols. Consumer tastes change, of course. But, over many years, a select number of symbols have been ubiquitous in the surveys.

Superman's shield is one of them. Through its soon-to-be one-hundred-year history, Superman's logo has undergone a considerable evolution. The Man of Steel's emblem has been tampered with around twenty-five times, every tweak echoing each era's evolving design and taste sensibilities. The origins of Superman's logo hark back to his first appearance in *Action Comics #1* in 1938. Initially, his emblem was just an "S" with no shield. The modified shield version of his emblem didn't evolve until 1940, a design that stayed unchanged until the 1970s, when the emblem once more bore a major overhaul. Even though the tweaks have been mostly aesthetic, they have nonetheless helped keep the shield contemporary and vital, guaranteeing that it still resonates with consumers of all ages. And, through all the changes, the shield's essential message is the same. The core "S" is a visual and powerful metaphor synonymous with Superman's strength, heroism, and resolute courage.

The Spiderman logo has been another constant in consumer polls of the most iconic and recognizable superhero symbols. As the alter-ego of high school student Peter Parker, Spiderman first appeared in *Amazing Fantasy #15* in 1962. He quickly became a consumer favorite, earning his own comic series the following year. As with Superman's shield, Spiderman's logo has undergone an evolution that has been closely coupled with the development of the character, again echoing the evolving times and tastes of comic book consumers.

The original Spiderman logo was pretty basic, if not a little dull. It was simply the word "Spiderman" in red lettering set out boldly against a blue background. In the late 1960s, artist John Romita Sr. took over the series and created a more dynamic and stylized version of Spiderman. This new look was of a more muscular and athletic-looking superhero. The costume was sleeker and more streamlined. And a new logo was made to complement the new look: an emblem of a large, red spider set against a field of black. Romita's design innovation quickly caught on with fans and became the standard for Spiderman logos until the late 1990s and early 2000s when the logo went through an accelerating pace of tweaks and modifications, again echoing the changing tastes of the times. Tweaks in shape, size, and color of the logo, as well as the addition of graphic elements like webs and shadows, helped evolve the emblematic core elements of Spiderman.

Finally, of course, there's Batman. This legendary figure from American consumer culture first appeared in *Detective Comics #27* in 1939 under the creative vision of writer Bill Finger and artist Bob Kane. With his dark and brooding demeanor, razor-sharp mind, and high-tech gadgetry, Batman managed to stand out even in a genre where brightly clad and muscle-bound superheroes prevail. Consequently, he became a firm consumer favorite, evolving into a cultural icon with a global following.

Arguably one of the two most iconic superheroes of all time, notwithstanding the DC/ Marvel debate, the Batman logo sits alongside Superman's shield as the most recognizable superhero logos around the globe. Needless to say, Batman is a complex character. As an alterego for the mega-wealthy Bruce Wayne, Batman's emblem needed to portray not just the strength of the wearer's wit and intelligence, but also how he was more than mere human, when the need arose, so that he could face up to foe with superhuman levels of power.

With Batman, the logo has an added backstory to spice up the legend. The choice and persona of the bat logo derives from a tale of Bruce Wayne as a kid. Having fallen into a dark crevice/hole/cave, Bruce was suddenly terrified of a colony of bats that careered and flapped around him in the darkness. The experience was so defining that, when an adult, Bruce chose the bat symbol to signify the fact he had overcome past phobias and vulnerabilities, and become a new person. The original Batman emblem showed the black stylized silhouette of a bat over a yellow background. The same symbol was then used as a Bat-Signal, the hazard warning sign that appears in the various readings of the Batman mythos. Typically, the signal appears as a modified searchlight, a stylized bat-symbol attached to the light so that a huge Bat emblem is projected onto the cloudy "sky" above Gotham City. In the tales of the Batman, the signal is used by the Gotham City Police Department as a means of summoning Batman in the event of significant peril, or as a means of psychological intimidation to the various villains of Gotham City.

Before we go on to explore the early twenty-first-century renaissance in the superhero genre, let's first consider the associated obsession with branding in late twentieth century capitalism.

NO LOGO

In 1999, Canadian author Naomi Klein wrote *No Logo*, a bestseller that brilliantly critiqued the power of the super-brands and super-logos. Klein's book dropped not just on the cusp of the new millennium, but also at the start of a new phase of globalization. Global brands like Microsoft and Shell, Pepsi and Coca-Cola, Disney and McDonald's,

Shell, Starbucks, and Nike were becoming hugely powerful corporations, buttressing up against local laws and civic opposition in the pursuit of ever increasing profits. The progress seemed relentless, as these waves of western capitalism crashed against the shores of the developing world, leaving in their wake human and environmental concerns, which were considered secondary to the needs of the corporations.

No Logo charted the irresistible rise of youth-oriented, style-savvy consumer capitalism, a new breed in which corporations flogged a romanticized lifestyle, rather than their mere physical product. With the actual production facilities moved out of sight (and out of mind), the corporations could center their operations in Europe and North America on increasingly serpentine and invasive marketing campaigns, whilst safe-guarding their brand through censorship and legal action. To take just a couple of examples, Disney once sued a small-town creche for painting an unauthorized mural of their characters, and as Naomi Klein observed in *No Logo*:

> McDonald's, meanwhile, continues busily to harass small shopkeepers and restaurateurs of Scottish descent for that nationality's uncompetitive predisposition toward the Mc prefix on its surnames. The company sued the McAlan's sausage stand in Denmark; the Scottish-themed sandwich shop McMunchies in Buckinghamshire; went after Elizabeth McCaughey's McCoffee shop in the San Francisco Bay Area; and waged a twenty-six-year battle against a man named Ronald McDonald whose McDonald's Family Restaurant in a tiny town in Illinois had been around since 1956.

Bitingly witty and polemical in places, *No Logo* won much respect for its tireless research and reportage, from the blunt and often nauseating declarations of the bull-shitting brand gurus to the personal reporting from "export processing zones," a.k.a. the sweatshops in the developing world's semi-lawless regions. For those who advocated the philosophy at the core of *No Logo*'s worldview, it became clear that corporations were becoming more powerful than governments. The worldview went something like this: time was when we the people had focused on the oppressive straitjackets of militarism, racism, or patriarchy. Now it was the super-brand corporations that were synonymous with all that was ill in the world. The fall of the Berlin Wall only made matters worse. The super-brands roamed across even more of the globe unfettered, and their brand-first mantra started to dominate party politics too.

No Logo shook the world of publishing. Klein's book became a bestseller in numerous countries and was translated into over thirty languages, with more than a million copies in print worldwide. Not bad for a twenty-nine-year-old writer previously unknown outside her native Canada. The swift success rocketed Klein to global fame and her subsequent books became bestsellers. And *No Logo* inspired many artists and musicians. For example, British rock band Radiohead were so inspired by Klein's book that they toured Europe in a tent to help escape from corporate-sponsored venues. Radiohead also considered naming their 2000 album *No Logo*, before finally settling on *Kid A*. Indeed, *No Logo* was such a success that Klein's publisher suggested, apparently oblivious to the irony, that they should copyright the book's title and logo. Still, others even proposed a *No Logo* clothing line.

IS CAPITALISM MALE AND PENETRATIVE?

Can we identify some way in which this rise of corporate power should perpetuate the dominance of male superheroes, both in terms of branding and profit? This idea that corporate capitalism is male, or somehow has gendered characteristics, is a subject of analysis and debate in the fields of political and social sciences. Of course, it's clear that, as an economic system, capitalism is not inherently gendered. It's neither male nor female. Yet, the ways in which it has been historically practiced, and continues to be practiced, along with its consequences for the planet, certainly can be interpreted as having gendered implications.

Let's look at some ways in which capitalism has been critiqued for having gendered effects or consequences. First up is the gender wage gap. In many capitalist countries, there exists a gender wage gap where women, on average, earn less than men for the same work. This is the outcome of historical discrimination and gendered labor divisions. And, it could be argued, we see a reflection of this in the assumption that male superheroes are more important than their female counterparts. What is Supergirl other than a two-decades-later female afterthought based on Superman? What is She-Hulk other than a woman who received an emergency blood transfusion from her cousin, Bruce Banner, and got a milder version of his Hulk condition? And what is Batgirl, no matter if it's Barbara Gordon, Stephanie Brown, or Cassandra Cain, other than a copycat version of Batman?

Another critique of the gendered effects of corporate capitalism is the division of labor. Capitalism has most often reinforced conventional gender roles, with men being more likely to hold high-paying positions in the workforce, while women have disproportionately occupied lower-paid and unpaid caring jobs. This has been traditionally reinforced by the consumerism and gender stereotypes of capitalism. Advertising and marketing most often perpetuate gender stereotypes to sell brands and products, reinforcing conventional notions of masculinity and femininity.

Over its long history, the realm of superheroes, in whatever media, has promoted stereotypes that communicate which gender, appearances, and behaviors are acceptable in society. For example, the scholarly study, *Gender Differences in Movie Superheroes' Roles, Appearances, and Violence*, produced a content analysis of 147 superheroes in eighty movies, and found that male heroes appeared much more frequently than female heroes.

Other gendered stereotypes include the fact that females are often portrayed as more likely to work in a group while males more likely to work alone. With regard to conventional notions of masculinity and femininity in the movies, males are more powerful, muscular, violent, and evil, while women are more attractive, thin, sexy/seductive, innocent, afraid, and helpless. When it comes to clothing, both costumes and non-costumes, you won't be surprised to find that females' clothes are more revealing on both the upper and lower bodies. Even though both genders often have special powers and weapons, male superheroes are more likely than females to have more than one special power and use more than one weapon. Male superheroes more often have super strength and resistance to injury, while females more often are able to manipulate elements, such as fire. Male superheroes are significantly more likely to use fighting skills, fire/flame weapons, and guns than females. The study concluded that, collectively, these movie depictions communicate conventional ideas about women as nurturing and fertile, passive and non-agentic background characters, and (white) men as powerful masters of reason who operate in the foreground.

Another important critique of the gendered effects of corporate capitalism is the environmental impact. Many critics suggest that capitalism's relentless pursuit of profit leads to the exploitation of natural resources and environmental degradation, which can disproportionately affect vulnerable populations, including women and marginalized communities. This question of the environmental impact of capitalism and how to fix it arose in an interview between British podcaster, Russell Brand, and Naomi Klein.

Naomi Klein: In order for us to rise to this challenge [of climate crisis], we have to let go of some of our crutches. There are stories that we tell ourselves to comfort ourselves that everything is really going to be okay . . . one is that the billionaires are going to save us, at the very last minute. Our oligarchs who have gotten so incredibly rich.

Russell Brand: Our Batman?

Naomi Klein: Well, our Batman appears to be Bill Gates. So there's been a few of these superhero billionaire stories where guys like Gates and [Richard] Branson have positioned themselves as, you know, they're gonna come in with a techno-fix at the last minute.

Russell Brand: Why the *last* minute?! For drama?

Naomi Klein: Because that's what superheroes do!

British spy writer John Le Carré is another author who has commented about corporate "superhero" power:

A ludicrous notion took root that we are saddled with to this day . . . It holds to its bosom the conviction that, whatever vast commercial corporations do in the short term, they are ultimately motivated by ethical concerns, and their influence upon the world is therefore beneficial. And anyone who thinks otherwise is a neo-Communist heretic . . . In the name of this theory, we look on apparently helpless while rainforests are wrecked to the tune of millions of square miles every year, native agricultural communities are systematically deprived of their livelihoods, uprooted and made homeless, protesters are hanged and shot, the loveliest corners of the world are invaded and desecrated, and tropical paradises are turned into rotting wastelands with sprawling, disease-ridden megacities at their center.

It's important to recognize that these gendered aspects of capitalism are not inherent to capitalism itself but are shaped by the historical, cultural, and social contexts in which capitalism operates. Efforts to address these gendered issues often involve changes in policies, regulations, and societal attitudes to promote gender equality within a capitalist framework. Additionally, alternative economic systems and models, such as socialism or cooperative economies, have been proposed as ways to address some of the gender disparities associated with capitalism.

HOW THE SUPERHERO BRANDS CAME TO RULE

With the burgeoning popularity of superhero symbols, what explains the late twentieth century/early twenty-first century renaissance in superhero films? The main factors fostering such popularity have undoubtedly shifted in recent years. And they've become increasingly tied to franchising and special effects, and less tied to their social and political relevance (just think of the social and political relevance of Captain America and Wonder Woman during WWII). With the new tech available, increasingly corporate film studios were more able to create vividly realistic fantasy worlds. And audiences were more willingly taken on the rollercoaster ride of the latest blockbuster movie, without feeling the need for the movies to be socially and politically relevant.

The situation is not totally black and white. Social and political contexts still imbue superhero movies to some extent. But the contextual relevance is no longer demanded by the vast majority of consuming moviegoers. Witness the increasingly apolitical and asocial movies in the superhero catalog. Indeed, some audiences, particularly *international* audiences, appear to prefer a *reduced* amount of American political contextualization, as it might be seen to make the movies *more* relevant in the context of foreign societies. However, a new question dawned: How far can corporate film studios peddle special effects–burdened films before they become anodyne, monotonous, and dull? Will consuming moviegoers continue to spend their money to see sequel after sequel?

MARKETING MARVEL

The year 1999 not only produced Klein's *No Logo* but was also a watershed year in the realm of superhero comics. Marvel was in deep difficulty. They needed cash, badly. But they believed superhero movies to be a tired trope. They believed the genre-defining superhero films, from Christopher Reeve's *Superman* of 1978 through Tim Burton's *Batman* of 1989 to the critically panned and campy *Batman & Robin* of 1997, were a dying breed, so Marvel sold off the movie rights of their most iconic heroes. First went Spiderman, sold to Sony for a bargain $7 million. (Not a bad investment by Sony when you consider that, as of June 2023, the Spiderman franchise had become a glorious box-office goldmine for Sony, coining in a cumulative global gross of around $8.5 billion over the span of ten films, and that figure doesn't even include the additional $1.5 billion spawned by the associated Sony Spiderman Universe films.) Shortly after the Spiderman sale, the X-Men brand was sold to 20th Century Studios. Together, the Spiderman and X-Men franchises would go on to reenergize the superhero film industry, both, as we will soon see, sitting in the top ten highest grossing movie franchises of all time.

Marvel's forecast that consumers were weary of superhero films proved to be dramatically wrong. And, as the Spiderman and X-Men franchises made stratospheric profits, Marvel needed to make their own mark on the silver blockbuster screen, but *without* using their most iconic characters. The likes of Spiderman, Wolverine, Cyclops, and Magneto were all off-limits thanks to the premature deal done with Fox and Sony.

WHAT MARVEL DID NEXT

How did Marvel recover? Let's remind ourselves just how much of a recovery they pulled off. As of 2022, the Marvel Cinematic Universe makes up the highest-grossing movie series, even when adjusted for inflation, surpassing J. K. Rowling's Wizarding World (eleven movies), Star Wars (twelve movies), Ian Fleming's James Bond (twenty-seven movies), and J. R. R. Tolkien's Middle-Earth (six movies) series:

1. Marvel Cinematic Universe
2. Spider-Man
3. Star Wars
4. Wizarding World
5. James Bond
6. Avengers
7. Fast & Furious
8. Batman
9. DC Extended Universe
10. X-Men

Marvel's astounding success has not come exclusively from their big-name characters. For example, Ant-Man, a superhero most consumers had probably never even heard of, earned $150 million more than the first Captain America film. And Marvel's success has contrasted sharply with their main competitor, the DC Cinematic Universe. Despite being the stable of history's most famously logo'd superheroes like Batman and Superman, DC has not enjoyed anywhere near the kind of cinematic success as Marvel. Why? Marvel believed its best marketing tactic was to grow its overall brand. Then, once that strong brand was established, hitherto relatively unknown characters could come to the fore and be made more compellingly visible and profitable. Naturally, tooling up a long-term tactic often brings its own problems and trials. And, in Marvel's case, the major trial was the challenge of fast money quickly versus the potential of more money later.

Marvel could easily have tossed some of their better-known characters into *The Avengers* and made a quick steal. Sure, these characters would not have had the iconic cachet of Superman and Batman, but consumers might still be keen to watch a pantheon of heroes on the silver screen. Rather, Marvel wisely chose to slowly build their brand, gradually growing the profile of individual heroes while trailing and teasing the plots and moving pictures to come. Marvel mapped its movies out until 2028, and that helped make their success all the more likely.

MARVEL PHASE ONE

Stage one of Marvel's cinematic plan was to make movies for individual heroes, and so it began with the four male Avengers: Captain America, Iron Man, the Hulk, and Thor. Once the buzz was created about these heroes, *The Avengers* was able to draw a far larger audience of consumers. Marvel may well have been aware of the cinematic legacy left by Akira Kurosawa's 1954 legendary Japanese film, *Seven Samurai* (initially released in the US as *The Magnificent Seven*). *Seven Samurai* was not just a marvelous movie. It was also the catalyst for a spin-off subgenre of movies that has flowed through cinema ever since. Arguably, *Seven Samurai* is the first film in which a team is assembled to carry out a mission. *Seven Samurai* certainly gave birth to its direct Hollywood remake, *The Magnificent Seven*, also to 1967's box office blockbuster,

The Dirty Dozen, about a real-life WWII unit behind enemy lines, and, finally, greatly influenced Marvel's *Avengers*.

As a result of Kurosawa's well-known legacy, the-gang's-back-together-again films like *The Avengers* or *Batman v Superman* get an almost immediate thumbs-up from consumers. They know this trope well. What Marvel knew, and DC didn't, was that the lure of such legacies was significant. Let's compare like with like. DC released *Batman v Superman* without independent features for its integral star superheroes. Likewise, DC's *Suicide Squad* had among its cast of heroes and villains maybe the all-time best villain in the Joker. But DC made the rookie mistake of introducing the whole suicide squad of characters in a single film. And *Justice League* confirmed that DC *still* hadn't learned from these errors. Individual movies allow consumers more time to mull over a character's backstory and their personal connection with the audience. As a result of Marvel's groundwork, consumers were already sold on each character by the time *The Avengers* hit the cinemas. Consequently, *The Avengers* took $1.5 billion at the box office to become the fifth highest-grossing film of all time.

Marvel's strategy was all about building the brand. They tapped into multiple channels. The cumulative branding of the first five Marvel Cinematic Universe (MCU) movies fed into *The Avengers*, which doubly grabbed the attention of consumers from the get-go. Marvel was patient. They didn't go for the big sale straightaway. Sure, the five movies leading up to *The Avengers* were still designed to make money. But there was more. They were designed to make the impact of *The Avengers* that much more dramatic. And the overall effect of their brand-building strategy not only resulted in the huge popularity of *The Avengers*, but also led to the further explosion of Marvel's global brand awareness. And this meant huge box office sales for subsequent movies and more revenue from the merchandising of all MCU characters. In short, Marvel nurtured and seduced their consuming audience before pushing for the big sale.

MARVEL'S OTHER PHASES

After all these years, *The Avengers* is still the tenth highest-grossing film of all time. In the wake of the success of *The Avengers*, and with their brand at peak market prominence, Marvel knew they had achieved mission success: the MCU had huge untapped potential. Now was the hour to cash in on their operation; continue to build established iconic characters, while bringing in lesser-known characters across other forms of media. And so, in cinema theaters, each Avenger got a further individual film to enhance their characters and capitalize on their consumer popularity. Meanwhile, a lesser-known superhero, Ant-Man, marched into box offices for a cool global $519 million.

On the totem pole of MCU, Ant-Man is pretty much way down the superhero hierarchy, yet the new power of Marvel's brand propelled the Ant-Man movie to its box office success. Consumers didn't *need* to know a detailed Ant-Man backstory to be persuaded to see the latest Marvel movie. The strength and sheer momentum of the MCU brand not only propelled *Ant-Man* forward, but also meant that the movie out-grossed *Captain America: The First Avenger* and rivaled the revenue of the 2008 *Iron Man* movie.

Beyond the movie theaters, Marvel got its tentacles into television. Starting with shows like *Agents of SHIELD, Agent Carter, The Inhumans,* and *Damage Control*, Marvel continued to flesh out the Avengers' story by showcasing minor characters. Moreover, they expanded into *Netflix* with yet another superhero hit in *Daredevil*. And the huge success of *Daredevil* triggered a suite of Netflix shows centered on lesser-known heroes of the MCU. And yet *Daredevil, Luke Cage* and *Jessica Jones* (an actual female superhero!) all received strong reviews, leading to further superhero shows such as *The Punisher, Iron Fist, WandaVision, Loki, She-Hulk, Secret Invasion,* and so on.

Movie	Release Date
Phase One	
Iron Man	May 2, 2008
The Incredible Hulk	June 13, 2008
Iron Man 2	May 7, 2010
Thor	May 6, 2011
Captain America: First Avenger	July 22, 2011
The Avengers	May 4, 2012
Phase Two	
Iron Man 3	May 3, 2013
Thor: The Dark World	November 8, 2013
Captain America: Winter Soldier	April 4, 2014
Guardians of the Galaxy	August 1, 2014
Avengers: Age of Ultron	May 1, 2015
Ant-Man	July 17, 2015
Phase Three	
Captain America: Civil War	May 6, 2016
Doctor Strange	November 4, 2016
Guardians of the Galaxy Vol. 2	May 5, 2017
Spider-Man: Homecoming	July 7, 2017
Thor: Ragnarok	November 3, 2017
Black Panther	February 16, 2018
Avengers: Infinity War	April 27, 2018
Ant-Man and the Wasp	July 6, 2018
Captain Marvel	March 8, 2019
Avengers: Endgame	April 26, 2019
Phase Four	
Black Widow	July 9, 2021
Shang-Chi and the Legend of the Ten Rings	September 3, 2021
Eternals	November 5, 2021

Spider-Man: No Way Home	December 17, 2021
Doctor Strange in the Multiverse of Madness	May 6, 2022
Thor: Love and Thunder	July 8, 2022
Black Panther: Wakanda Forever	November 11, 2022
Phase Five	
Ant-Man and the Wasp: Quantumania	February 17, 2023
Guardians of the Galaxy Vol. 3	May 5, 2023
The Marvels	November 10, 2023
Deadpool 3	May 3, 2024
Captain America: Brave New World	July 26, 2024
Thunderbolts	December 20, 2024
Blade	February 14, 2025
Phase Six	
Fantastic Four	May 2, 2025
Avengers: The Kang Dynasty	May 1, 2026
Avengers: Secret Wars	May 7, 2027

Table 2. List of Marvel Cinematic Universe movies. The MCU is the shared Universe in which all the films are set. The movies from phase one through phase three are known as "The Infinity Saga." The movies from phase four through phase six are known as "The Multiverse Saga."

Series	Season	Episodes	Original Release
ABC Series			
	1	22	September 24, 2013
	2	22	September 23, 2014
	3	22	September 29, 2015
Agents of S.H.I.E.L.D.	4	22	September 20, 2016
	5	22	December 1, 2017
	6	13	May 10, 2019

(Continued on next page)

	7	13	May 27, 2020
Agent Carter	1	8	January 6, 2015
	2	10	January 19, 2016
Inhumans	1	8	September 29, 2017
Netflix Series			
	1	13	April 10, 2015
Daredevil	2	13	March 18, 2016
	3	13	October 19, 2018
	1	13	November 20, 2015
Jessica Jones	2	13	March 8, 2018
	3	13	June 14, 2019
Luke Cage	1	13	September 30, 2016
	2	13	June 22, 2018
Iron Fist	1	13	March 17, 2017
	2	10	September 7, 2018
The Defenders	1	8	August 18, 2017
The Punisher	1	13	November 17, 2017
	2	13	January 18, 2019
Young Adult Series			
	1	10	November 21, 2017
Runaways	2	13	December 21, 2018
	3	10	December 13, 2019
Cloak & Dagger	1	13	June 7, 2018
	2	13	April 4, 2019
Adventure Into Fear Series			
Helstrom	1	10	October 16, 2020

Table 3. *List of Marvel Cinematic Universe television series. The MCU first expanded to television on the creation of Marvel Television in 2010, with that studio producing twelve series with ABC Studios. The main ABC series were inspired by the films and featured movie characters, while Netflix produced connected crossover series.*

BUT WHERE ARE ALL THE FEMALE SUPERHEROES?

Given the triumph and momentum of Marvel's marketing strategy, and the ease with which they were able to successfully and profitably fit in "lesser" heroes of their canon into their theater program, why not take such a golden opportunity to showcase a few female superheroes? In short, if it works for Ant-Man, why wouldn't it work for superwomen? Let's look at some compelling evidence.

Take the example of Jane Foster in the *Thor* movies. When Natalie Portman was first approached to play the role, she reacted with consummate professionalism and intelligence:

> I signed on to do it before there was a script. And Ken [Branagh], who's amazing, who is so incredible, was like, "You can really help create this character." I got to read all of these biographies of female scientists like Rosalind Franklin who actually discovered the DNA double helix but didn't get the credit for it. The struggles they had and the way that they thought—I was like, "What a great opportunity, in a very big movie that is going to be seen by a lot of people, to have a woman as a scientist." She's a very serious scientist. Because in the comic she's a nurse and now they made her an astrophysicist. Really, I know it sounds silly, but it is those little things that make girls think it's possible. It doesn't give them a [role] model of "Oh, I just have to dress cute in movies."

Yet, the initial reaction to Portman's casting was that she was too petite and generally not what "people" imagined the character to be. For "people" here, one should actually read "young, white, and male," as traditionally that's the demographic of the established audience. Ten months down the line, after Portman had dutifully engaged in a regime of intensive workouts and a high-protein diet to bulk up, the same people applauded Portman for arms that looked like they could "actually throw giant hammers at baddies' heads."

With such a sophisticated analysis, it's so easy to see the "problem" with female superheroes. It seems not to matter that the proportion of female consumers of the MCU has increased. When the prospect of a more feminist portrayal of a superwoman arises, one that would challenge the pervasive masculinity of the genre, a "problem" is flagged by marketing (not for nothing did American comedian Bill Hicks darkly joke about the "evil machinations" of "marketing people!").

What does this mean for the way in which moviemakers depict female superheroes of the future? For now, the answer appears to be that writers and directors within the genre may have to try subverting some gender stereotypes while tolerating others. In other words, moviemakers allow a merely token female representation in order to avoid losing consumers, audiences, and profits. For example, while Natalie Portman became tolerably muscular, her superwoman character was still subservient to Chris Hemsworth's *Thor* by focusing on the fact that she is first and foremost Thor's love interest.

Let's be quite clear. We've come a long way from the original portrayal of hypersexualized female superheroes. And the MCU franchise has at least tried to cast female leads and support women's issues. For instance, Marvel's standalone movie *Black Widow* was partially conceived to speak to the discourse around the #TimesUp and #MeToo movements. And *Thor: Love and Thunder* explores the value of female friendships, with female actor Tessa Thompson confessing that her character Valkyrie was "happy to have found a new sister."

So there's little doubt that female consumers can relate to powerful superwomen and their tales and, consequently, develop positive attitudes to the MCU in the process. And *that* means more superhero genre movies need to be made with the female consumer in mind. As we have seen from tables 2 and 3 above, however, such movies are few and far between. After all, it took Marvel a decade to give Black Widow her own movie after her initial introduction in 2010's *Iron Man 2*. And so it's easy to see why many critics say that the movies of MCU still portray women as ancillaries: old-school damsels in distress, romantic interests, or subservient in some other way to their male leads.

Indeed, actor Scarlett Johansson was highly critical of the earlier hypersexualization of her Black Widow character in *Iron Man 2*. She told Collider in 2021, "You look back at *Iron Man 2* and while it was really fun and had a lot of great moments in it, the character is so sexualized, you know? Really talked about like she's a piece of something, like a possession or a thing or whatever." Talking about her character's evolution, Johansson said she was treated like "a piece of ass" at first: "Tony [Stark, played by Robert Downey Jr.] even refers to her as something like that at one point. What does he say? 'I want some.' Yeah, and at one point calls her a piece of meat and maybe at that time that actually felt like a compliment. You know what I mean?" Johansson went on to say that her self-worth was "probably measured against that type of comment," with regard to her physical appearance, a decade back, but she's since evolved:

> I'm more accepting of myself, I think. All of that is related to that move away from the kind of hypersexualization of this character . . . Now people, young girls, are getting a much more positive message, but it's been incredible to be a part of that shift and be able to come out the other side and be a part of that old story, but also progress. Evolve.

Likewise, Scarlet Witch, arguably one of the most powerful characters in the MCU, is often defined by her relationships with the males in her life. For example, in *Doctor Strange in the Multiverse of Madness*, Scarlet

Witch is seen to typify many unfavorable female tropes, including the "hysterical woman" and "monstrous mother."

THE HYPERSEXUALIZED STEREOTYPE

The truth of the matter is, even though some superwomen are portrayed as powerful female characters, a closer examination of, for example, the three MCU superheroines, Black Widow, Scarlet Witch, and Mystique, shows that they are constantly dogged with difficulties and problems that supermen never have to face. This subordination and dependency may mollify male consumers that superwomen aren't a real threat to the macho undertow of the genre, but it does little for the aspirations of female consumers.

In one Canadian study out of Brock University, most women when asked to comment on superhero graphic novels and movies said they disliked and avoided the DC Comics character of Catwoman because she was depicted as Machiavellian and emotional. Another study, out of London Metropolitan University, found that exposure to movie messages of powerlessness can lead girls to feel demoralized and dissatisfied with their own identities, and that the overly sexualized representation of superwomen can result in lower self-esteem in women.

Concurrently, some rebel against sexual stereotypes. *The Hawkeye Initiative*, a parodic *Tumblr* account that posts wry comments on the depiction and treatment of female characters and superheroes in comic books, satirizes the male-dominated gaze of the sequential art genre by depicting men in the same absurd costumes and poses normally reserved for female characters.

BOX OFFICE BITE BACK?

Naturally, if more movies and sequential art were being made *by* women *for* women, the challenge of having to battle against such tokenistic cinematic depictions would be less likely to arise in the first place. Marvel has dismissed criticism of its superwomen. Marvel Studios president Kevin Feige told the press at an event in Los Angeles in 2015 that "our films have been full of smart, intelligent, powerful women." The topic arose while Feige was promoting *Ant-Man*, in which the female character, Hope van Dyne, is prevented from performing heroic actions by her protective father, Hank Pym. The reporter conducting the interview asked why female superheroes had been sidelined in most of the studio's movies. Feige replied that the decision to make a film starring a female hero was not a reactionary move in response to an outcry from fans and the media, but part of Marvel's strategy to go for "the powerful woman versus the damsel in distress," pointing to the then-recent release of female-led superwomen films and TV programs like *She-Hulk* and *Ms. Marvel*.

The narrative of Black Widow in *Avengers: Age of Ultron* had been a topic of debate earlier that year after Chris Evans and Jeremy Renner joked in an interview that the character was a "slut" due to her romantic relationships with both Captain America and The Hulk. Consumers were also disturbed by a scene in the movie in which Black Widow recalled the horror of being forcibly sterilized. In the scene, she referred to herself as a monster because of it, which outraged many moviegoers. Feige commented:

In terms of essays written about Black Widow in *Ultron*, I think they're all valid. Everybody's opinions are valid. [. . .] To suggest that female characters can't have multiple dimensions is also ludicrous. That Black Widow went through a program in which she was forced to have her reproductive organs removed is probably a little upsetting to her. So that people would be upset that she's upset—that's a little strange.

No doubt Marvel has a job on their hands trying to keep everyone happy. The studio allegedly felt a backlash from conservative male fans in response to a supposed feminist agenda directing the studio's program. The 2019 *Captain Marvel* movie, for instance, was said to be bringing feminism to the MCU, yet poor reviews and consumer ratings were explained, at least in part, by the deemed political correctness and a story centered on female agency.

French philosopher Simone de Beauvoir once wrote that "representation of the world, like the world itself, is the work of men; they describe it from their own point of view, which they confuse with absolute truth." Although fictional, representations of superwomen matter, as they help construct new ideations of female identity, the acceptance of which can then promote changed behaviors. Men are capable of creating complex and imperfect female characters, but it requires more than good intentions.

Many female scholars contend that the dominance of men in the superhero genre means that many female consumers are left feeling alienated and powerless to change the stereotypes, precisely as they're not perceived to be the target audience. And so, without more and better female movie superheroes telling women's tales, the male-dominated genres will continue to alienate female consumers and also fall short of their creative and commercial potential. If our media consumer culture allows that superheroes exist as mythological versions of our shared reality as complicated, flawed, and noble humans, there's no reason it can't also admit that some of them are women.

INDEX